Sustainable Development Goals Series

The **Sustainable Development Goals Series** is Springer Nature's inaugural cross-imprint book series that addresses and supports the United Nations' seventeen Sustainable Development Goals. The series fosters comprehensive research focused on these global targets and endeavours to address some of society's greatest grand challenges. The SDGs are inherently multidisciplinary, and they bring people working across different fields together and working towards a common goal. In this spirit, the Sustainable Development Goals series is the first at Springer Nature to publish books under both the Springer and Palgrave Macmillan imprints, bringing the strengths of our imprints together.

The Sustainable Development Goals Series is organized into eighteen subseries: one subseries based around each of the seventeen respective Sustainable Development Goals, and an eighteenth subseries, "Connecting the Goals", which serves as a home for volumes addressing multiple goals or studying the SDGs as a whole. Each subseries is guided by an expert Subseries Advisor with years or decades of experience studying and addressing core components of their respective Goal.

The SDG Series has a remit as broad as the SDGs themselves, and contributions are welcome from scientists, academics, policymakers, and researchers working in fields related to any of the seventeen goals. If you are interested in contributing a monograph or curated volume to the series, please contact the Publishers: Zachary Romano [Springer; zachary.romano@springer.com] and Rachael Ballard [Palgrave Macmillan; rachael.ballard@palgrave.com].

Clio Ciaschini · Luca Salvati

Sustainable Development

Official Statistics and Thematic Indicators as a Compass to Understand Complexity

Springer

Clio Ciaschini
Department of Social
and Economic Sciences
Polytechnic University of Marche
Ancona, Italy

Luca Salvati
Faculty of Economics
Sapienza University of Rome
Rome, Italy

ISSN 2523-3084 ISSN 2523-3092 (electronic)
Sustainable Development Goals Series
ISBN 978-3-032-12558-3 ISBN 978-3-032-12559-0 (eBook)
https://doi.org/10.1007/978-3-032-12559-0

This Springer imprint is published by the registered company Springer Nature Switzerland AG
The registered company address is: Gewerbestrasse 11, 6330 Cham, Switzerland

Contents

The Challenge of Sustainable Development Indicators

1

Introduction

Sustainable Development (SD) has been routinely defined as a 'development that meets the needs of the present without compromising the ability of future generations to satisfy their own'. This operational definition qualifies the notion of 'sustainability' as a target oriented towards a dynamic balance between man and ecosystems (Panagiotopoulos et al., 2024). At the same time, SD refers to the holistic approach (and the underlying spatio-temporal processes) leading to sustainability principles and conditions (Ruggerio, 2021). Within this context, two cases emerged, *i.e.* 'weak' or 'strong' sustainability, connected with a sustainable exploitation of natural resources (Ekins, 2014). Following the paradigm of weak sustainability, the desirable development model should ensure a stock of (at least) non-decreasing human and natural resources until the next generation (Dufrénot, 2023).

In economic terms, the concept of 'sustainability' assumes the full substitutability between the two types of capital; a decrease in one capital corresponds to an increase in the other capital, maintaining the sum of human and natural capitals constant over time (Ekins et al., 2003). The principle of 'strong sustainability' puts this statement under discussion (Sachs, 2015). Following the Brundtland Report, SD paradigm has been consolidated, becoming a hegemonic concept in sustainability science—being progressively associated with economic, social, and environmental sustainability (Hajian & Kashani, 2021). As a basic definition, SD is fair and inclusive, practically achievable, and therefore liveable and accessible to all population segments (Šlaus & Jacobs, 2011), being in turn balanced across time and space (Hariram et al., 2023).

SD connects geographical and ecological themes (Vardopoulos et al., 2020, 2023a, 2023b) with demographic aspects, as well as the sectors (and dimensions) of culture (Vardopoulos et al., 2023a, 2023b), technology (Vardopoulos et al., 2024), and institutions (Acosta Castellanos et al., 2021). In this sense, SD has been regarded, in the latter decades, as the organizational principle to attain human development objectives supporting economies and societies (e.g. Lessmann & Rauschmayer, 2013). The immediate result of a long term development path oriented to sustainability is a dynamic state characteristic of any society (Bendell, 2022; Ben-Eli, 2018; Doukas et al., 2023).

In this perspective, sustainability is a concept that, even if vague, clarifies that quantifiable limits to growth exist. It is also defined as a political call to action (Mensah, 2019). The concept evolved throughout time, combining scientific and normative approaches (e.g. Salvati & Zitti, 2009), starting from the definition of the Brundtland Report to reach a more holistic and integrated vision, concretised in the 2030 Agenda

C. Ciaschini and L. Salvati, *Sustainable Development*, Sustainable Development Goals Series,
https://doi.org/10.1007/978-3-032-12559-0_1

(Vardopoulos & Karytsas, 2019). In 2012, Rio + 20 marked a relevant step in achieving a global SD path, dealing with issues such as the fight against poverty, energy, and water (Doukas et al., 2024a, 2024b). Starting from the seminal contributions to this conference, a new working group developed the Sustainable Development Goals, delineating 17 main objectives and designing 169 targets declared in 2015 within the 2030 Agenda (Ribeiro Hoffmann et al., 2024).

Contemporaneously, with Paris COP21 in 2015, a global agreement was signed with the aim of containing global warming to less than 2 °C compared to pre-industrial levels. To encourage a balanced and sustainable global development, the 2030 Agenda further redefined the sustainable development goals, including all world countries, thus overcoming the distinction between advanced and emerging economies. In September 2015, the 193 Countries attending the Summit of the United Nations adopted the 2030 Agenda for Sustainable Development, through a specific resolution of the General Assembly (A/RES/70/1). The negotiations occurred contextually to those of the United Nations on funding of national and regional development, which determines financial means to actuate the development agenda itself (sensu Arrow et al., 2004). The United Nations resolution of 2015 recognized the unsustainability of the present ecological, economic, and social development model overcoming the idea that sustainability is primarily a matter for ecologists (e.g. Francaviglia et al., 2019).

The Role of Indicators in Sustainable Development

A proper study of Sustainable Development could not prescind from the recognition of its indicators as crucial quantitative tools to measure and monitor the environmental, social and economic performances of territorial and social systems. Indicators, as green income (namely, GDP), greenhouse gas emissions, the per-capita water availability, the coverage on healthcare and educational services or the share of energy from renewable sources, are intended to evaluate the capability of development systems to preserve and increase natural, human and social capitals (Dufrénot, 2023). They provide objective data, comparable over time and between different geographical contexts, representing crucial tools to define quantifiable limits to growth and elaborate policies effectively acknowledging objectives of 2030 Agenda for Sustainable Development (Ekins et al., 2003; Vardopoulos & Karytsas, 2019). Nevertheless, these indicators seem not to accomplish the objective of capturing the complexity of socio-ecological processes. They should instead be complemented by qualitative and quantitative (mixed) approaches and subjective indicators based on perceptions and individual experiences of social actors (Šlaus & Jacobs, 2011), aimed at integrating and estimating dimensions such as life quality, social cohesion, equity perception, resiliency of local communities, and trust in institutions (Hariram et al., 2023). Such approaches include surveys based on self-report, semi-structured interviews, focus groups, techniques of citizen science, perception maps and participative evaluations, to collect information directly from community and stakeholders.

Moreover, methods relying on social media and digital platforms effectively monitor opinions and collective feelings on sustainability topics, thus providing dynamic indicators on public perceptions (Ben-Eli, 2018). A joint consideration of objective data and subjective information can enable more comprehensive multidimensional evaluations, capable of identifying both criticalities and potentials within socio-ecological systems, thereby supporting the design of effective, targeted policies that address hidden inequalities and assess the social acceptability of economic and environmental measures. Since combining the empirical measures of processes with experiences and perceptions of the involved actors, this approach facilitates the design of dedicated and effective policies, thus enabling a more participated, inclusive and resilient decisional process (Doukas et al., 2023; Ruggerio, 2021; Salvati & Zitti, 2009).

Within this framework, 'indicator dashboards' reveal as a crucial tool for ex-ante monitoring, ex-post evaluation, and communication at all operational and stakeholder levels. This tool further allows to clearly aggregate, display and compare several indicators (Panagiotopoulos et al., 2024), thus facilitating data interpretation and the communication of empirical results within policymakers, citizens and scientific environments (Vardopoulos et al., 2020). Beyond this descriptive role, dashboards could be included into Integrated Decision-Support Systems, (IDSS), combining quantitative and qualitative information, historical data, predictive models, policy scenarios and impact simulations (Ben-Eli, 2018). These advanced systems enable an effective evaluation of trade-offs, identify synergies and estimate long-term effects, thus transforming data in proactive strategic management tools to support more informed, transparent and adaptive decisions (Mensah, 2019).

In conclusion, an effective selection and adoption of indicators should respect the principles of equity and inclusion at the basis of Sustainable Development. Indicators should measure effectively the key variables of economic, social and environmental systems, also accounting for territorial inequalities and social disparities, thus ensuring a balanced sustainability accessible to the whole community (Hajian & Kashani, 2021). In addition, integrating objective indicators with subjective and advanced decision systems may support the achievement of a quantifiable, participative and adaptive Sustainable Development model, able to respect available natural and social resources and promote a balanced, resilient and long-lasting growth (Ekins, 2014; Sachs, 2015).

'One Health' and Sustainable Development

In recent years, the concept of Sustainable Development, (SD), emerged as the leading principle of global policies, basically aimed at ensuring a dynamic equilibrium between economic growth, social equity and environmental safeguard. The operative definition retrieved in the Bruntland Report (1987)—"a development that meets the needs of the present without compromising the ability of future generations to meet their own needs"—provides the basis for a holistic vision of sustainability, intended as equilibrium between men and ecosystems. This perspective has been progressively enriched by interdisciplinary approaches widening the field of application of the concept, including cultural, institutional and technological aspects (Ruggerio, 2021). Among others, healthcare emerges as transversal and essential dimension of sustainable development. COVID-19 pandemics has highlighted the fragile interconnections between human health, ecosystems and productive systems, requiring renewed growth models and paradigms of global governance (Vardopoulos et al., 2023a, 2023b).

Within this framework, 'One Health' approach, recognized and promoted by the main Agencies of United Nations, e.g. FAO, UNEP, WHO e WOAH, constitutes an integrated framework to face emerging risks originated by the interaction between man, animals, plants and environment (Samhouri et al., 2025). The One Health paradigm is based on the recognition of the interconnection between the different components of the biosphere and on the necessary connection of the human healthcare with the animal and environmental ones (Pitt & Gunn, 2024). According to the definition proposed by ONU and its technical organs, One Health represents a holistic and integrated approach aimed at optimizing and balancing these dimensions to prevent healthcare threats, promote sustainable development, and strengthen the resilience of socio-ecological systems (Beever & Morar, 2019). Therefore, the One Health approach complements the traditional paradigm of Sustainable Development, focussing on healthcare as key variable in the planning and evaluation processes (Mumford et al., 2023). The integration between healthcare, environmental and socioeconomic indicators suggest the design of new monitoring tools underling the complex interactions between man and the environment (Caiado et al., 2018).

In this direction, the design of integrated indicators, based on both quantitative data and qualitative perceptions (e.g. Mayoux & Chambers, 2005), represents a necessary condition to identify criticalities, hidden inequalities and systemic risks hardly captured by traditional models (Burgass et al., 2017). The connection between healthcare and Sustainable Development implies the combination of objective and subjective indicators (Lyytimäki et al., 2020). Quantitative metrics, as greenhouse gas emissions, levels of atmospheric pollution, biodiversity, or the rates of avoidable mortality, provide an essential basis to measure the progresses towards sustainability (Movilla-Pateiro et al., 2021; Tanzil & Beloff, 2006; Zhang et al., 2008). Nevertheless, these measures should be complemented by qualitative and participative approaches to capture the perceived and experienced dimensions of welfare and sustainability (Moldan et al., 2012).

The integration of 'One Health' paradigm within the sustainable development strategies has significant implications for the structures of governance and elaboration of public policies (Lee & Brumme, 2013). It requires an intersectoral approach to overcome the administrative fragmentation between areas as healthcare, agriculture, environmental safeguard, energy and territorial planning (Buse & Hawkes, 2015). Coherently with the European healthcare approach in policy designs, this paradigm privileges the policy coherence and the coordinate management of risks (Faijue et al., 2024). Under the operative viewpoint, 'One Health' implementation imposes the construction of integrated surveillance systems collecting and sharing epidemiological, environmental and biological data. These systems should ensure interoperability, transparency and timeliness, thus requiring the cooperation between national and international organizations. In this perspective, contextual mechanisms of inclusive participation, involving citizens, territorial entities, local communities, forms and organizations of civil society, allow to integrate local and scientific knowledges, generating more adaptive and contextually relevant strategies (Jeleff et al., 2022).

Under the economic and territorial profile, the 'One Health' approach supports the definition of resilient development strategies, to reduce the systemic vulnerability and improve the response capability to multiple healthcare, climatic or socioeconomic crises. Within this perspective, besides an isolated sector and cost, healthcare is considered an integrated capital (human, natural and social) to be preserved and enhanced in the long run (Tambo et al., 2017). In conclusion, the integration between 'One Health' and Sustainable Development delineates a new architecture of global governance based on cooperation, transparency and shared responsibility. This novel 'structure' allows to face the challenges of the XXI century, from climate change to socioeconomic inequalities, until the food and healthcare safety, through preventive interdisciplinary and participatory instruments, addressing a general target such as the 'total' welfare of people, communities and ecosystems (Teshome, 2024).

Envisaging Sustainability as a Practical Notion

The 2030 Agenda entitled 'Transforming our world: the 2030 Agenda for Sustainable Development' is made of 92 paragraphs delineating 17 sustainable development objectives 169 associated objectives and 232 associated indicators. The Agenda adopts an integrated SD vision involving all countries, without distinguishing between advanced and emerging economies, with the aim of defining national strategies to attain SDGs. The Agenda promotes a global (integrated) picture to face economic, social, and environmental problems, including issues such as poverty, climate change, gender equality, healthcare and energy. In summary, the 2030 Agenda widens the concept of sustainability concerning the Millennium Development Goals (MDGs), adopting an integrated approach that may effectively underline the interconnection of global problems (Fig. 1.1). The practical implementation of Agenda 2030 requires the contribution of the whole society, firms and the public sector, from universities to civil society, using an approach coordinated by the United Nations.

The 169 objectives connected to the 17 SDGs (Fig. 1.2) face issues such as poverty, healthcare, education, climate, gender equality, urbanization, and energy. The 2030 Agenda, with respect to the previous Millennium Development Goals, promotes an integrated paradigm well interconnected with sustainability, acting as an operative

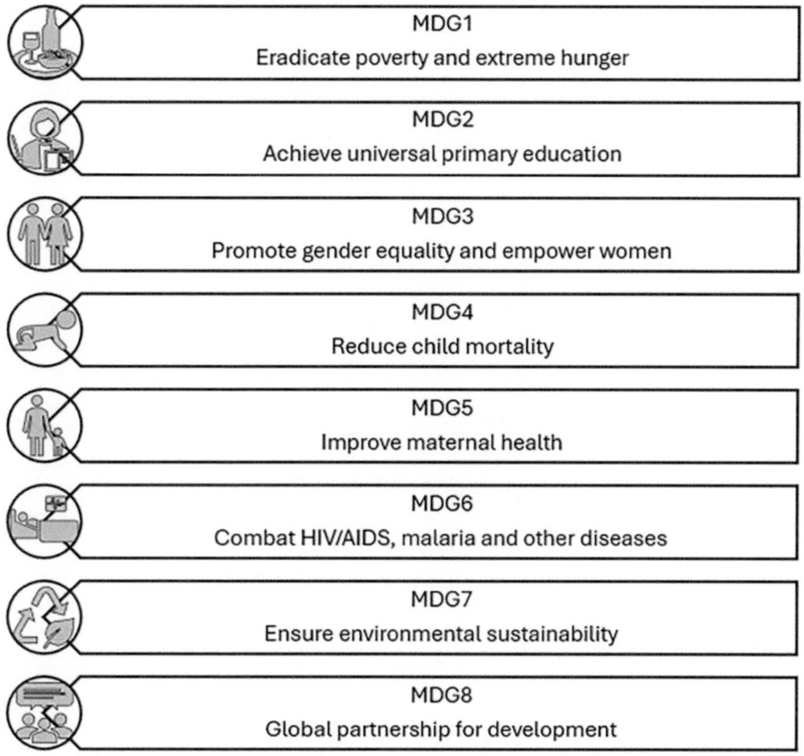

Fig. 1.1 A graphical illustration of the eight MDGs of United Nations. *Source* Authors' elaboration of United Nations sustainable development strategies

'umbrella' for global development. The action plan is defined using the five 'P's, namely:

(i) People (eliminate poverty and ensure decent living conditions);
(ii) Prosperity (economic and environmental welfare);
(iii) Peace (design policies favouring peace);
(iv) Partnership (interaction within countries to a common and coordinated achievement of the Goals);
(v) Planet (actions to protect natural environment and climate).

The resolution of objectives consists of a wide inter-government agreement constituting the Post 2015 Development Agenda.

SDGs substitute and reinforce the Millennium Development Goals (MDGs) established in 2000 and not completely fulfilled, adopting an integrated approach that underlines the interconnection of global issues with local dimensions (Halisçelik & Soytas, 2019). Differently from MDGs—addressing specifically the target of developing countries—SDGs refer to all countries worldwide and have been designed and approved synergistically by governments of all countries. The attainment of SDGs was (and still is) evaluated continuously through a wide set of more than 230 indicators. Within this framework, the so-called 'custodians' of SDGs, collaborate with statistical offices to elaborate and refine the methodologies shaping the indicators reported in United Nations Statistics (Global SDG database). Among the different organizations, the World Bank is one of the main custodians, responsible for 22 indicators, quantifying different aspects related to crucial topics connected with SD (Fig. 1.3). Within this context, the reduction of poverty to less than 3 per cent by 2030, and the promotion of prosperity favouring increases in income of the poorest 40 per cent of the population, also correspond to the twin goals set by the World Bank Group in 2013 (Huan et al., 2021).

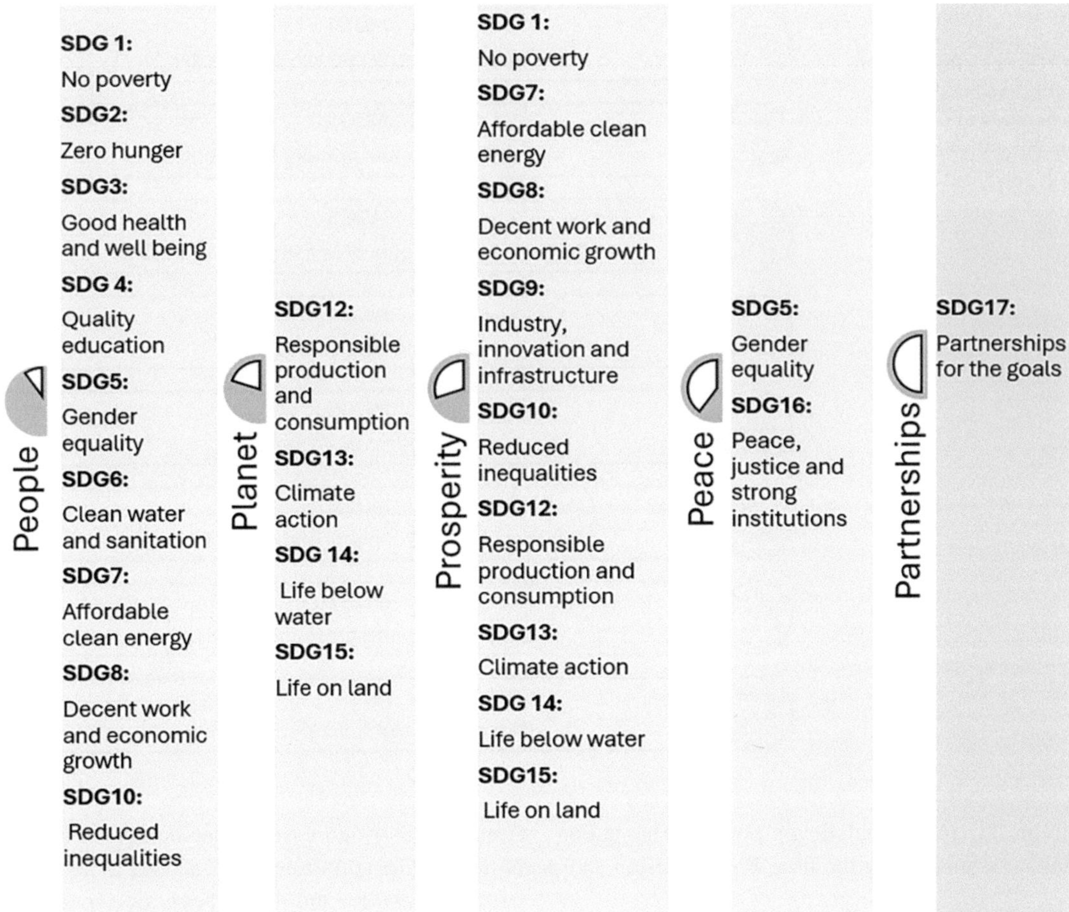

Fig. 1.2 A graphical illustration of the five P's of the United Nations' SDGs. *Source* Authors' elaboration of United Nations strategies on sustainable development

Sustainable Development Goals

In Europe, the strategy of the European Commission for sustainable development has increasingly aimed at achieving a prosperous, cleaner and equitable, continent. Within this context, the European Community, for many years, has acted as a leader of the path towards a global strategy of sustainability. For instance, inclusive economic growth requires progress within the set of complementary targets, following a specific SD objective (namely, SDG8) related to e.g. workers' rights (Kavouras et al., 2022). Decent work is both a relevant result and a leading factor within the agenda 'centred on man' and aimed at inclusive growth, warranting that each member of the society could participate in the creation of economic value and benefit of growth advantages. Following this vision, SDG8 of the 2030 Agenda ('decent working conditions and economic growth') is well-oriented to the issue of social sustainability, specifically focussing on human and labour rights (Cernev & Fenner, 2020). Sustained economic growth was assumed to derive from a virtuous circle involving transformative changes in economics and society with multiple channels leading the dynamic process to the objectives of SDG8 (Van Niekerk, 2024).

Practical experiences demonstrate that economic growth cannot be sustained when it is mainly based on the circular demand, export of natural resources, or, simply, the accumula-

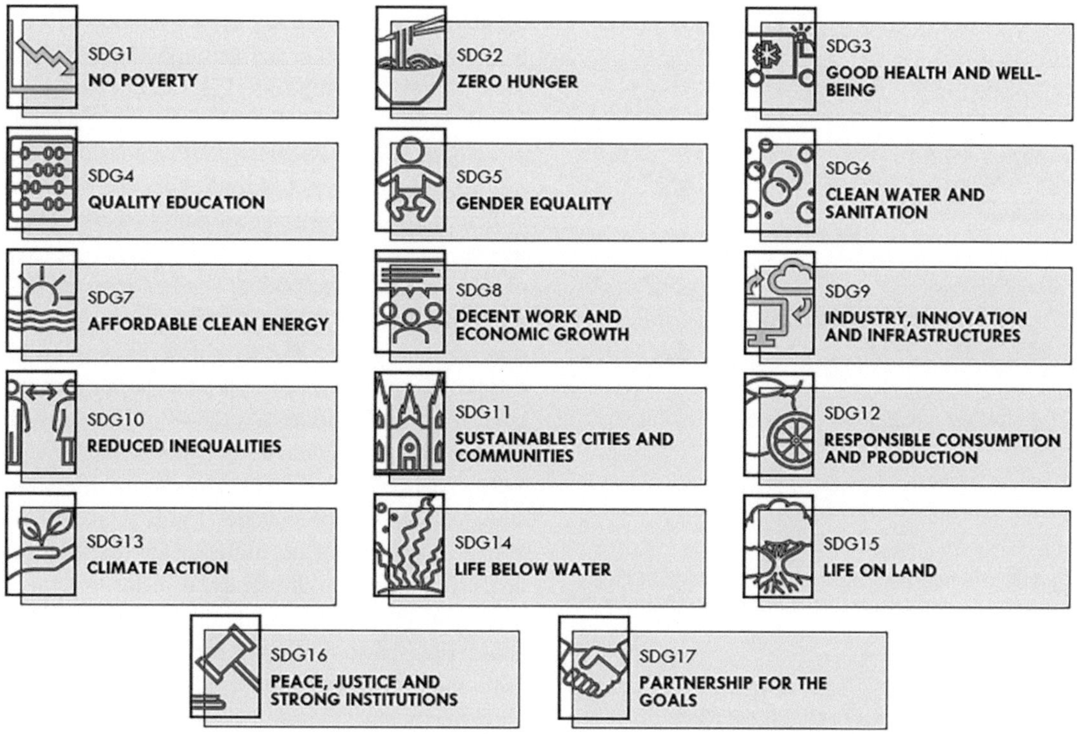

Fig. 1.3 A graphical illustration of the 17 United Nations SDGs. *Source* Authors' elaboration of UN strategies on sustainable development

tion of production factors and monetary capita (Paun et al., 2019). A sustained growth—balanced in all its technical, cultural and operational components—instead requires structural changes, inherent transformation in the productive base, industrial diversification and service innovation to support the long-term growth of factor productivity, income and levels/opportunities of decent work (Cavalcanti et al., 2011; Kocornik-Mina et al., 2021; Pothen & Welsch, 2019). Structural transformations in turn derive from the enrichment and empowerment of the operational knowledge of a given society (Fouré et al., 2013). As a consequence, progress on SDG10 ('reduced inequalities') is crucial to warrant that productivity increases are translated into higher incomes and salaries (Girdzijauskas et al., 2022). In addition, increasing learning opportunities in the workplace, empowering the disadvantaged worker categories, and inclusive growth together with decent work, improve the diversity and depth of the basic knowledge of a country and the whole economy (Kabeer, 2021). To nurture human capital through learning and experience, the virtuous cycle centred on SDG8 and SDG10 further requires progresses on multiple complementary objectives on the base of different SDGs (Hosono, 2022), as illustrated in Fig. 1.4.

As previously remarked, SDG9 refers to 'quality and resilient 'infrastructures' and 'industrialisation' (Vardopoulos et al., 2024), SDG7 relates to 'clean and trustable energy provision' speeding structural transformation (Vardopoulos et al., 2023a, 2023b), SDG1 specifically to 'social protection', SDG4 to 'education' (Moustairas et al., 2022), SDG5 to 'gender responsibility', while SDG16 mainly refers to 'quality institutions'. Objectives of SDG8—and of other sustainable development targets cited before—are interconnected with the objectives of the ILO (International Labour Organisation). The reference to 'full employment and decent work for everyone' borrows the terminology of ILO Work Agenda, testifying how sustainable devel-

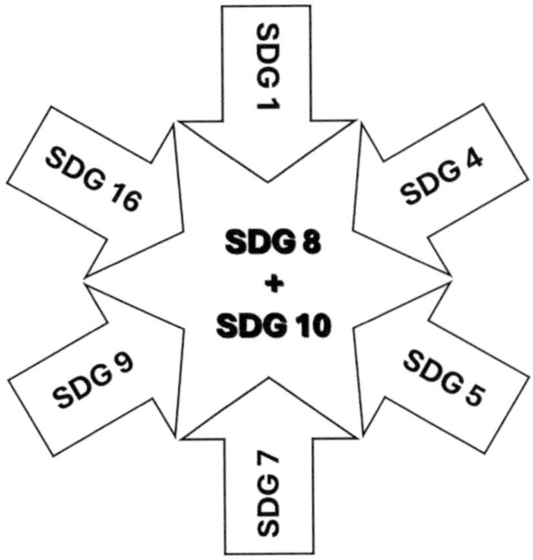

Fig. 1.4 A summary representation of relevant SDGs for economic growth. *Source* Authors' elaboration of UN strategies on sustainable development

opment and labour dignity have been interconnected topics and faced with the same operative plan for more than 20 years. A renowned ILO Declaration for the Future of Work (adopted since the International Conference of Work in June 2019) borrows the formulation of SDG8 to 'promote a sustained, sustainable, and inclusive economic growth' –reflecting an anthropocentric approach centred on the future of work. Underlining the needs of a given society, this strategy opened a virtuous path towards sustainability and economic development, accounting for the environmental dimension, and emphasizing the importance of a complete transition towards individual behaviours of responsible consumption.

Economic, Social and Environmental Indicators of Sustainable Development

Based on these premises, the present book proposes an in-depth analysis of spatio-temporal trends of paradigmatic sustainability indicators released by the World Bank, as related to the main areas of interest for SD science (representative of economic, environmental and social issues), as well as policy topics connected with the maintaining of peace through the establishment of solid institutions (Fig. 1.5). These indicators were specifically selected as covering, directly or indirectly, all the SDGs, possibly from multiple perspectives (Ahrens et al., 2025). These indicators have been investigated continuously over a period covering about 60 years (approximately from 1960 to 2020). The few missing data recorded in the collection of quantitative information released by the World Bank have been estimated to ensure comparisons among countries (e.g. Sallustio et al., 2018). The analysis consists of a visual data description through maps, followed by a descriptive analysis of the main statistics at the global scale (Egidi et al., 2020). When appropriate, a multivariate exploratory analysis completes the study, offering graphical plots that act as a visual aid to better understand the spatio-temporal dynamics associated with individual indicators, adopting a country-wide perspective (Vardopoulos et al., 2021).

The Rationale Behind the Selection of Indicators

The selection of indicators presented in this volume responds to the necessity to develop a coherent, balanced and scientifically sound picture for the analysis of sustainable development in its three (economic, social and environmental) composing dimensions. The aim is that of providing interpretative tools to describe and evaluate sustainability as a dynamic and integrated process, in line with the objectives of the 2030 Agenda and with the evolution of recent international literature. The chosen indicators represent key components of contemporary development processes, aimed at estimating the interdependencies between economic growth, social welfare and environmental quality. The adopted approach reflects a systemic version, in which sustainability emerges as a result of the interactions between human, natural, and institutional capital, rather than a simple static equilibrium (Valentinov, 2021). Technically, the selection of indicators is based on statistical robustness criteria, interna-

Income growth (1, 2, 3, 8) **Age dependency ratio** (3, 5, 8) **Adolescent fertility rate** (2,3,4,5,8,16)	**Life expectancy at birth, female** (1, 2, 3, 4, 5, 10) **CO_2 emissions** (7, 9, 11, 13)	**Agricultural land** (6, 12, 13, 14, 15) **Forest rents** (7, 13, 14, 15) **Military expense** (8, 9, 16)

Fig. 1.5 Selected indicators representative of global-scale SDGs. *Source* authors' elaboration of World Bank indicators and UN strategies of sustainable development

tional comparability and temporal availability, to ensure the replicability of the empirical analysis and the coherence of achieved results.

Indicators derive from official sources, as the World Bank, the United Nations, and the OECD, thus ensuring data reliability and their comparability with the adopted quantitative methodologies. Among several (potentially available) indicators, those offering a complementary information have been preferred since they reduce semantic overlaps and analytical duplications, thus ensuring a better readability of results. The proposed indicators should clarify the latent relationship between economy, society and environment, and support the identification of criticalities and synergies characteristic of contemporary territorial systems (Lehtonen, 2004). Within this framework, the combination of economic (e.g. income growth and military expenditure), social (e.g. life expectancy and adolescent fertility rate), and environmental (e.g. greenhouse gas emission or the share of agriculture and forest area), makes 'sustainability' estimation an integrated and multidimensional concept (Sartorius, 2006).

Economic and Social Development in a 'Sustainability' Perspective

In recent years, SDGs have acquired relevance as alternative (and more comprehensive) indicators of social and economic development (Dědeček & Dudzich, 2022). These circumstances have been qualified in a new concept representing the so-called 'beyond GDP' perspective (Yang et al., 2020), requiring more effective measures of economic development (Giovannini & Rondinella, 2018), rather than a mere assessment of the economic output of a given country (Ivković, 2016). SDGs act as multi-scale indicators, applicable from local to global levels, approaching several topics together, from poverty to social inequalities, from climate change to economic growth and urban issues (Blair et al., 2021).

This debate has been carefully considered in Chap. 2, examining the long-term expansion of incomes across the world. By examining wealth and population dynamics that affect the attainment of almost all SDGs, among which SDG1 and SDG8, and going beyond the simple economic dimension (Dribe & Scalone, 2014), the indicator deals with the containment of poverty within a global context of migration flows and demographic change (González-Leonardo & Spijker, 2023). An increasing population also highlights several issues concerning crop intensification and efficiency in food distribution (Perales & Bernard, 2023). The chapter also refers extensively to the empirical works of Redclift (2005) and Spaiser et al. This concept, further detailed in subsequent chapters of this book, casts doubt on the association of 'per-capita GDP' to SDG8 (sustainable development growth), going beyond the strictly economic aspects (Coscieme et al., 2020; Meadows et al., 1972; Raworth, 2017; Turner, 2014).

Nevertheless, as pointed out in Chap. 2, even addressing exclusively economic aspects, the indicator of 'per-capita GDP', seems to be effective in depicting the globally heterogeneous paths of economic growth (Capello & Lenzi, 2014), as well as the persisting inequalities within countries (Doukas et al., 2024a, 2024b) and world regions (Chen et al., 2014). In connection with the country-level economic output, the age dependency ratio, in Chap. 3, quantifies the ratio of dependents on the working–age population (Hondroyiannis & Papapetrou, 2001), allowing the study of changes in population composition and proportions between working and not-working-age classes (Hammer et al., 2015). This indicator informs on issues connected to SDG4 on education, usually improved in dependent age classes (Hock & Weil, 2012). When coupled with the fertility rate, the indicator also supports the study on the advancement in pursuing gender equality (SDG5) and reducing working inequalities (SDG10). The ultimate goal should be that of decent working conditions and sustainable economic growth (SDG8).

The pursued socioeconomic development cannot be decoupled from an operational investigation of the demographic aspects connected with economic growth (Alemayehu et al., 2010). Within this context, Chap. 4 examines empirical evidence related to adolescent fertility rates, an indicator reflecting improved maternal health and a reduced infant mortality rate (Kearney & Levine, 2012). Pregnant young unmarried women, indeed, undergo harder conditions in reproductive healthcare services, influencing the attainment of SDG3 on good health and well-being (Singh, 1998). An early pregnancy could also prevent women in the achievement of personal (and professional) goals. Adolescent mothers are less likely to reach higher levels of education when matching family and work responsibilities (Hotz et al., 1997), hindering the attainment of SDG4 on quality education, in turn affecting working conditions and, ultimately, SDG8 (economic growth). Early pregnancy also implies wide barriers in the challenging path to gender equality, e.g. SDG5, and reduction of inequalities in working environments and other contexts, e.g. SDG10 (Morris, 2023).

The specific spatio-temporal trends in adolescent fertility rates go along with life expectancy at birth, as described in Chap. 5, since maternal health challenged by an early pregnancy, strongly impacts the life expectancy of the new-borns. A specific section, within SDG3, addresses this issue, i.e. SDG3.1—containing preventable deaths of new-borns and children under 5 years as well as maternal mortality. Within this context, even if with important divides between advanced countries and emerging economies (Aburto et al., 2022; Kashyap, 2022; Perrin et al., 2022), MDG4 ('reduce child mortality') to be achieved within 2015, remains, up to now, not fulfilled. Within the UN 2030 Agenda, the containment of child mortality rates was addressed via multiple targets, e.g. SDG1 on poverty, SDG3 on healthcare, and SDGs 4, 5, and 8, all dealing with socioeconomic aspects (e.g. Benassi et al., 2020). In a worldwide, increasingly older population, the elderly acquire a relevant position among vulnerable groups and constitute active agents of social development (Acheampong et al., 2019; Bora & Saikia, 2018; Hossain et al., 2022). Policies promoting the well-being of older adults relate, on the one side, to the infrastructural interventions favouring elderly well–being, as SDG9 and SDG11, concerning innovative infrastructures and smart cities able to meet the requirement of an increasingly aged population. On the other hand, socioeconomic targets refer to SDG1 and SDG10, on fighting poverty and reducing inequalities, and to SDG3 on healthcare and well-being (Cheng et al., 2020).

Environmental Dynamics in a 'Sustainability' Perspective

A more specific framework dealing with environmental issues is provided in Chap. 6 when describing the empirical evidence stemming from a comprehensive and comparative analysis of CO_2 emissions (Vardopoulos & Konstantinou, 2017; Zhou et al., 2021). This indicator primarily refers to SDG7, pursuing clean energy (Georgiou et al., 2025), and SDG13, against climate change (Godil et al., 2021). The indicator also integrates a latent debate dealing with the risk of a progressive decoupling of SD and economic growth basically connected with SDG8 (Bouma et al., 2021; Khan et al., 2019; Li & Wang, 2023). To this purpose, Chap. 7, discusses the relevance of the agricultural sector in economic systems of both advanced and emerging countries. This sector is crucial in providing food and fibre (e.g. Tomao et al., 2017), directly responding to SDG12 on consumption and production (Wu et al., 2014; Ramankutty et al., 2018; Volkov & Fomin, 2018; Vardopoulos, 2025; Vardopoulos et al., 2025). Nevertheless, agricultural production affects CO_2 emissions, in turn resulting in climate change (Biasi et al., 2019), addressed within SDG13 (Bach et al., 2020). Actions concerning agricultural production also pertain to SDG15, 'life on land' (Bhatt, 2023). A more latent connection of this indicator with SDG1 ('zero poverty'), SDG2 ('zero hunger') and SDG8 ('sustainable environmental and socioeconomic development' of countries) is envisaged (e.g. Quaranta et al., 2020).

Positioning within the climate change issue (Ferrara et al., 2017), Chap. 8 investigates the spatio-temporal dynamics of forest rents as directly connected with SDGs concerning climate change (SDG 13) and life on land (SDG5), and indirectly connected with each of the remaining Sustainable Development Goals (D'Agata et al., 2024; Escrivà Saneugenio et al., 2024; Swamy et al., 2018). This indicator specifically highlights the relevance of natural resources in economic growth (Ferrara et al., 2016; Mancino et al., 2016; Sathaye et al., 2007), without neglecting the detrimental effects connected to an 'over the threshold' exploitation of wood resources (Jagger et al., 2019; Yu, 2023).

The persistent geo-political instability of recent times finally acquired a crucial role in challenging socioeconomic development targets (Doukas et al., 2025; Monaco, 2023). Country–level inequalities between affluent and emerging countries reflect a context in which conflict–affected environments achieve lower levels of SDGs than advanced economies (Feng et al., 2023; Le Caous & Huarng, 2020; Rao & Rao, 2019). A decade after the establishment of the 2030 SDG Agenda, robust and stable institutions favouring justice and peace, still not exist at a global level (Ihebuzor et al., 2020; McDermott et al., 2019; Silander, 2020). This context induces a consideration of the most effective way to attain a peaceful society, free from conflicts, in line with SDG16 (Wakelin, 2021). A first step in this direction could be that of quantifying the long-term expenditure on military equipment, as illustrated in Chap. 9. The desirable reduction in these expenditures contrasts with the present political context defined by (negative) strategic interactions between countries (Singh & Singh, 2024), often resulting in an item of increased expenditure on weapons. More constructive interactions should contribute to the creation of peaceful and inclusive societies (Aizenman & Glick, 2006).

Concluding Remarks

In general, SDGs should provide an effective framework to quantify and assess the progress in the attainment of global socioeconomic development (Ibn-Mohammed et al., 2021). They provide a multifaceted picture of all SD dimensions (Bali Swain & Yang-Wallentin, 2020), not only accounting for the economic pillar but also including more tricky issues such as some latent environmental processes and the increasingly pivotal role of climate change (Griggs et al., 2014). Nevertheless, the efforts directed at the attainment of 2030 SD targets, i.e. an economic growth able to preserve and equally distribute natural resources as well as human-produced goods and services, seem to be not completely effective (e.g. Dahl, 2012). A more likely consequence would in turn bring to the exhaustion of all the available natural and human derived resources (Boström, 2012). A continuous monitoring system based on a few pillars and thematic indicators, instead of a more extensive information dashboard that can be redundant and plethoric (Allen et al., 2020), is the operational proposal introduced in the present book. Such information, scrutinized year by year and country by country, may effectively provide the necessary operational base for any policy action, indicating advancements and critical issues that can be removed with appropriate strategic management of development paths.

Acknowledgements The authors warmly thank all the students of the 'Economic statistics' course (academic year 2021–2022) held by Luca Salvati in the second semester (from February to May 2022) at University of Macerata, Italy. Students in this class implemented a collective exercise with official statistics aimed at the construction of selected indicators of sustainable development using, among others, quantitative information from World Bank database. Together with the instructor, the students performed various statistical analyses and graphs that helped reading the complexity of recent development paths in both advanced economies and emerging countries. While all students participating to the exercise implemented novel analyses and an original reading of official data, we are particularly grateful to Alessandra Muzi, Marlene Corvaro, Sara Di Felice, Giuseppe Maltese, Serena Gabrielli and Elena Paniccià, for their invaluable production of nice and informative statistical elaborations, maps and charts illustrating the specific pattern of selected indicators over time and space.

References

Aburto, J. M., Schöley, J., Kashnitsky, I., Zhang, L., Rahal, C., Missov, T. I., & Kashyap, R. (2022). Quantifying impacts of the COVID-19 pandemic through life-expectancy losses: A population-level study of 29 countries. *International Journal of Epidemiology, 51*(1), 63–74.

Acheampong, M., Ejiofor, C., Salinas-Miranda, A., Wall, B., & Yu, Q. (2019). Priority setting towards achieving under-five mortality target in Africa in context of sustainable development goals: An ordinary least squares (OLS) analysis. *Global Health Research and Policy, 4*(1), 1–16.

Acosta Castellanos, P.M., Queiruga-Dios, A., & Álvarez, L.G. (2021). Inclusion of education for sustainable development in environmental engineering. A systematic review. *Sustainability, 13*(18), 10180.

Ahrens, D., Benedikter, S., & Giessen, L. (2025). Rethinking synergies and trade-offs at the forest-sustainable development goals (SDGs) Nexus—A systematic review. *Sustainable Development.*

Aizenman, J., & Glick, R. (2006). Military expenditure, threats, and growth. *Journal of International Trade and Economic Development, 15*(2), 129–155.

Alemayehu, T., Haider, J., & Habte, D. (2010). Determinants of adolescent fertility in Ethiopia. *Ethiopian Journal of Health Development, 24*(1).

Allen, C., Reid, M., Thwaites, J., Glover, R., & Kestin, T. (2020). Assessing national progress and priorities for the Sustainable Development Goals (SDGs): Experience from Australia. *Sustainability Science, 15*, 521–538.

Arrow, K. J., Dasgupta, P., Goulder, L., Daily, G., Ehrlich, P. R., Heal, G. M., Levin, S., Maler, K.-G., Schneider, S., Starrett, D. A., & Walker, B. (2004). Are we consuming too much? *Journal of Economic Perspectives, 18*(3), 147–172.

Bach, E. M., Ramirez, K. S., Fraser, T. D., & Wall, D. H. (2020). Soil biodiversity integrates solutions for a sustainable future. *Sustainability, 12*(7), 2662.

Bali Swain, R., & Yang-Wallentin, F. (2020). Achieving sustainable development goals: Predicaments and strategies. *International Journal of Sustainable Development and World Ecology, 27*(2), 96–106.

Beever, J., & Morar, N. (2019). The epistemic and ethical onus of 'One Health.' *Bioethics, 33*(1), 185–194.

Benassi, F., Cividino, S., Cudlin, P., Alhuseen, A., Lamonica, G. R., & Salvati, L. (2020). Population trends and desertification risk in a Mediterranean region, 1861–2017. *Land Use Policy, 95*, Article 104626.

Bendell, J. (2022). Replacing sustainable development: Potential frameworks for international cooperation in an era of increasing crises and disasters. *Sustainability, 14*(13), 8185.

Ben-Eli, M. U. (2018). Sustainability: Definition and five core principles, a systems perspective. *Sustainability Science, 13*(5), 1337–1343.

Bhatt, R. P. (2023). Achievement of SDGS globally in biodiversity conservation and reduction of greenhouse gas emissions by using green energy and maintaining forest cover. *GSC Advanced Research and Reviews, 17*(3), 1–21.

Biasi, R., Brunori, E., Ferrara, C., & Salvati, L. (2019). Assessing impacts of climate change on phenology and quality traits of *Vitis vinifera* L.: The contribution of local knowledge. *Plants, 8*(5), 121.

Blair, M. J., Gagnon, B., Klain, A., & Kulišić, B. (2021). Contribution of biomass supply chains for bioenergy to sustainable development goals. *Land, 10*(2), 181.

Bora, J. K., & Saikia, N. (2018). Neonatal and under-five mortality rate in Indian districts with reference to Sustainable Development Goal 3: An analysis of the National Family Health Survey of India (NFHS), 2015–2016. *PLoS ONE, 13*(7), Article e0201125.

Boström, M. (2012). A missing pillar? Challenges in theorizing and practicing social sustainability: introduction to the special issue. *Sustainability: Science, Practice and Policy, 8*(1), 3–14.

Bouma, J., Pinto-Correia, T., & Veerman, C. (2021). Assessing the role of soils when developing sustainable agricultural production systems focused on achieving the UN-SDGs and the EU Green Deal. *Soil Systems, 5*(3), 56.

Burgass, M. J., Halpern, B. S., Nicholson, E., & Milner-Gulland, E. J. (2017). Navigating uncertainty in environmental composite indicators. *Ecological Indicators, 75*, 268–278.

Buse, K., & Hawkes, S. (2015). Health in the sustainable development goals: Ready for a paradigm shift? *Globalization and Health, 11*(1), 13.

Caiado, R. G. G., Leal Filho, W., Quelhas, O. L. G., de Mattos Nascimento, D. L., & Ávila, L. V. (2018). A literature-based review on potentials and constraints in the implementation of the sustainable development goals. *Journal of Cleaner Production, 198*, 1276–1288.

Capello, R., & Lenzi, C. (2014). Spatial heterogeneity in knowledge, innovation, and economic growth nexus: Conceptual reflections and empirical evidence. *Journal of Regional Science, 54*(2), 186–214.

Cavalcanti, T. V. D. V., Mohaddes, K., & Raissi, M. (2011). Growth, development and natural resources: New evidence using a heterogeneous panel analysis. *The Quarterly Review of Economics and Finance, 51*(4), 305–318.

Cernev, T., & Fenner, R. (2020). The importance of achieving foundational sustainable development goals in reducing global risk. *Futures, 115*, Article 102492.

Chen, M., Zhang, H., Liu, W., & Zhang, W. (2014). The global pattern of urbanization and economic growth: Evidence from the last three decades. *PLoS ONE, 9*(8), Article e103799.

Cheng, X., Yang, Y., Schwebel, D. C., Liu, Z., Li, L., Cheng, P., & Hu, G. (2020). Population ageing and mortality during 1990–2017: A global decomposition analysis. *PLoS Medicine, 17*(6), Article e1003138.

Coscieme, L., Mortensen, L. F., Anderson, S., Ward, J., Donohue, I., & Sutton, P. C. (2020). Going beyond gross domestic product as an indicator to bring coherence to the sustainable development goals. *Journal of Cleaner Production, 248*, Article 119232.

D'Agata, A., Cudlin, P., Vardopoulos, I., Schinaia, G., Corona, P., & Salvati, L. (2024). Assessing the spatial coherence of forest cover indicators from different data sources: A contribution to sustainable development reporting. *Ecological Indicators, 158*, Article 111498.

Dahl, A. L. (2012). Achievements and gaps in indicators for sustainability. *Ecological Indicators, 17*, 14–19.

Dědeček, R., & Dudzich, V. (2022). Exploring the limitations of GDP per capita as an indicator of economic development: A cross-country perspective. *Review of Economic Perspectives, 22*(3), 193–217.

Doukas, Y.E., Vardopoulos, I., & Petides, P. (2024). Challenging the status quo: a critical analysis of the common agricultural policy's shift towards sustainability. In *Climate change in regional perspective. European Union and Latin American initiatives, challenges, and solutions*. (pp. 115–134). Cham, Switzerland: United Nations University Series on Regionalism, Vol. 27, Springer Nature

Doukas, Y. E., Maravegias, N., Ioannides, S., Salvati, L., & Vardopoulos, I. (2025). Economic resilience and sustainability in the European Union: A comparative analysis of pre- and posteconomic crisis. *Environmental Sustainability and Global Change* (pp. 257–273). Elsevier.

Doukas, Y. E., Maravegias, N., Salvati, L., & Vardopoulos, I. (2024a). Resilienza economica e sostenibilità: Indicatori macroeconomici e dinamiche politiche pre- e post-crisi nell'Unione Europea. *Documenti Geografici, 3*, 491–525.

Doukas, Y. E., Salvati, L., & Vardopoulos, I. (2023). Unraveling the European agricultural policy sustainable development trajectory. *Land, 12*(9), 1749.

Dribe, M., & Scalone, F. (2014). Social class and net fertility before, during, and after the demographic transition: A micro-level analysis of Sweden 1880–1970. *Demographic Research, 30*, 429–464.

Dufrénot, G. (2023). New thinking on sustainable development and growth. *New Challenges for Macroeconomic Policies: Economic Growth, Sustainable Development, Fiscal and Monetary Policies* (pp. 147–191). Springer International Publishing.

Egidi, G., Salvati, L., & Vinci, S. (2020). The long way to tipperary: City size and worldwide urban population trends, 1950–2030. *Sustainable Cities and Society, 60*, Article 102148.

Ekins, P. (2014). Strong sustainability and critical natural capital. *Handbook of sustainable development* (pp. 55–71). Edward Elgar.

Ekins, P., Simon, S., Deutsch, L., Folke, C., & De Groot, R. (2003). A framework for the practical application of the concepts of critical natural capital and strong sustainability. *Ecological Economics, 44*(2–3), 165–185.

Escrivà Saneugenio, F., D'Agata, A., Vardopoulos, I., & Salvati, L. (2024). Afforestation-deforestation interplay as a driver for urban fringe landscape change and sustainability. In *Sprawl and the city: realizing a sustainable Mediterranean urbanization* (pp. 105–128). Hauppauge, NY, USA: Nova Science Publishers. ISBN: 9798895300060

Faijue, D. D., Segui, A. O., Shringarpure, K., Razavi, A., Hasan, N., Dar, O., & Manikam, L. (2024). Constructing a one health governance architecture: A systematic review and analysis of governance mechanisms for One Health. *European Journal of Public Health, 34*(6), 1086–1094.

Feng, Y., Hu, J., Afshan, S., Irfan, M., Hu, M., & Abbas, S. (2023). Bridging resource disparities for sustainable development: A comparative analysis of resource-rich and resource-scarce countries. *Resources Policy, 85*, Article 103981.

Ferrara, A., Kelly, C., Wilson, G. A., Nolè, A., Mancino, G., Bajocco, S., & Salvati, L. (2016). Shaping the role of 'fast' and 'slow' drivers of change in forest-shrubland socio-ecological systems. *Journal of Environmental Management, 169*, 155–166.

Ferrara, C., Carlucci, M., Grigoriadis, E., Corona, P., & Salvati, L. (2017). A comprehensive insight into the geography of forest cover in Italy: Exploring the importance of socioeconomic local contexts. *Forest Policy and Economics, 75*, 12–22.

Fouré, J., Bénassy-Quéré, A., & Fontagné, L. (2013). Modelling the world economy at the 2050 horizon. *Economics of Transition, 21*(4), 617–654.

Francaviglia, R., Di Bene, C., Farina, R., Salvati, L., & Vicente-Vicente, J. L. (2019). Assessing "4 per 1000" soil organic carbon storage rates under Mediterranean climate: A comprehensive data analysis. *Mitigation and Adaptation Strategies for Global Change, 24*, 795–818.

Georgiou, C., Maialetti, M., Zorpas, A. A., Salvati, L., & Vardopoulos, I. (2025). Policy and planning for "zero-net energy" real estate development in rural and exurban districts. *Environmental Sustainability and Global Change* (pp. 291–303). Elsevier.

Giovannini, E., & Rondinella, T. (2018). Going beyond GDP: theoretical approaches. In *Handbook of Research on Economic and Social Well-Being* (pp. 1–51). Chichester, Edward Elgar.

Girdzijauskas, S., Streimikiene, D., Griesiene, I., Mikalauskiene, A., & Kyriakopoulos, G. L. (2022). New approach to inflation phenomena to ensure sustainable economic growth. *Sustainability, 14*(1), 518.

Godil, D. I., Yu, Z., Sharif, A., Usman, R., & Khan, S. A. R. (2021). Investigate the role of technology innovation and renewable energy in reducing transport sector CO_2 emission in China: A path toward sustainable development. *Sustainable Development, 29*(4), 694–707.

González-Leonardo, M., & Spijker, J. (2023). The impact of Covid-19 on demographic components in Spain, 2020–31: A scenario approach. *Population Studies, 77*(3), 497–513.

Griggs, D., Smith, M.S., Rockström, J., Öhman, M.C., Gaffney, O., Glaser, G., & Shyamsundar, P. (2014). An integrated framework for sustainable development goals. *Ecology and Society, 19*(4).

Hajian, M., & Kashani, S.J. (2021). Evolution of the concept of sustainability. From Brundtland Report to sustainable development goals. In *Sustainable resource management* (pp. 1–24). Amsterdam: Elsevier.

Halisçelik, E., & Soytas, M. A. (2019). Sustainable development from millennium 2015 to sustainable development goals 2030. *Sustainable Development, 27*(4), 545–572.

Hammer, B., Prskawetz, A., & Freund, I. (2015). Production activities and economic dependency by age and gender in Europe: A cross-country comparison. *Journal of the Economics of Ageing, 5*, 86–97.

Hariram, N. P., Mekha, K. B., Suganthan, V., & Sudhakar, K. (2023). Sustainalism: An integrated socio-economic-environmental model to address sustainable development and sustainability. *Sustainability, 15*(13), 10682.

Hock, H., & Weil, D. N. (2012). On the dynamics of the age structure, dependency, and consumption. *Journal of Population Economics, 25*(3), 1019–1043.

Hondroyiannis, G., & Papapetrou, E. (2001). Demographic changes, labor effort and economic growth: Empirical evidence from Greece. *Journal of Policy Modeling, 23*(2), 169–188.

Hosono, A. (2022). Transforming economies for jobs and inclusive growth. *SDGs, Transformation, and Quality Growth: Insights from International Cooperation* (pp. 103–120). Springer.

Hossain, M. M., Abdulla, F., Banik, R., Yeasmin, S., & Rahman, A. (2022). Child marriage and its association with morbidity and mortality of under 5 years old children in Bangladesh. *PLoS ONE, 17*(2), Article e0262927.

Hotz, V. J., Klerman, J. A., & Willis, R. J. (1997). The economics of fertility in developed countries. *Handbook of Population and Family Economics, 1*, 275–347.

Huan, Y., Liang, T., Li, H., & Zhang, C. (2021). A systematic method for assessing progress of achieving sustainable development goals: A case study of 15 countries. *Science of the Total Environment, 752*, Article 141875.

Ibn-Mohammed, T., Mustapha, K. B., Godsell, J., Adamu, Z., Babatunde, K. A., Akintade, D. D., & Koh, S. C. L. (2021). A critical analysis of the impacts of COVID-19 on the global economy and ecosystems and opportunities for circular economy strategies. *Resources, Conservation and Recycling, 164*, Article 105169.

Ihebuzor, N., Lawrence, A. W., Tamunomiebi, M., Lawrence, D. O., & Lawrence, B. A. (2020). The impacts of violent conflicts, injustice and weak institutions on the effective delivery of the SDGs and agenda 2030. *International Journal of Innovative Research and Development, 9*(7), 95–102.

Ivković, A. F. (2016). Limitations of the GDP as a measure of progress and well-being. *Ekonomski Vjesnik/econviews–review of Contemporary Business, Entrepreneurship and Economic Issues, 29*(1), 257–272.

Jagger, P., Bailis, R., Dermawan, A., Kittner, N., & McCord, R. (2019). SDG 7: Affordable and clean energy–how access to affordable and clean energy affects forests and forest-based livelihoods. *Sustainable development goals: Their impacts on forests and people* (pp. 206–236). Cambridge University Press.

Jeleff, M., Lehner, L., Giles-Vernick, T., Dückers, M. L., Napier, A. D., Jirovsky-Platter, E., & Kutalek, R. (2022). Vulnerability and one health assessment approaches for infectious threats from a social science perspective: A systematic scoping review. *The Lancet Planetary Health, 6*(8), e682–e693.

Kabeer, N. (2021). Gender equality, inclusive growth, and labour markets. *Women's economic empowerment* (pp. 13–48). Routledge.

Kashyap, R. (2022). Quantifying impacts of the COVID-19 pandemic through life-expectancy losses: a population-level study of 29 countries. *International Journal of Epidemiology, 51*(1), 63–74).

Kavouras, S., Vardopoulos, I., Mitoula, R., Zorpas, A. A., & Kaldis, P. (2022). Occupational health and safety scope significance in achieving sustainability. *Sustainability, 14*(4), 2424.

Kearney, M. S., & Levine, P. B. (2012). Why is the teen birth rate in the United States so high and why does it matter? *Journal of Economic Perspectives, 26*(2), 141–166.

Khan, S. A. R. (2019). The nexus between carbon emissions, poverty, economic growth, and logistics operations-empirical evidence from Southeast Asian countries. *Environmental Science and Pollution Research, 26*(13), 13210–13220.

Kocornik-Mina, A., Bastida-Vialcanet, R., & Eguiguren Huerta, M. (2021). Social impact of value-based banking: Best practises and a continuity framework. *Sustainability, 13*(14), 7681.

Le Caous, E., & Huarng, F. (2020). Economic complexity and the mediating effects of income inequality: Reaching sustainable development in developing countries. *Sustainability, 12*(5), 2089.

Lee, K., & Brumme, Z. L. (2013). Operationalizing the One Health approach: The global governance challenges. *Health Policy and Planning, 28*(7), 778–785.

Lehtonen, M. (2004). The environmental–social interface of sustainable development: Capabilities, social capital, institutions. *Ecological Economics, 49*(2), 199–214.

Lessmann, O., & Rauschmayer, F. (2013). Re-conceptualizing sustainable development on the basis of the capability approach: A model and its difficulties. *Journal of Human Development and Capabilities, 14*(1), 95–114.

Li, R., & Wang, Q. (2023). Does renewable energy reduce per capita carbon emissions and per capita ecological footprint? New evidence from 130 countries. *Energy Strategy Reviews, 49*, Article 101121.

Lyytimäki, J., Salo, H., Lepenies, R., Büttner, L., & Mustajoki, J. (2020). Risks of producing and using indicators of sustainable development goals. *Sustainable Development, 28*(6), 1528–1538.

Mancino, G., Nolè, A., Salvati, L., & Ferrara, A. (2016). In-between forest expansion and cropland decline: A revised USLE model for soil erosion risk under land-use change in a Mediterranean region. *Ecological Indicators, 71*, 544–550.

Mayoux, L., & Chambers, R. (2005). Reversing the paradigm: Quantification, participatory methods and pro-poor impact assessment. *Journal of International Development, 17*(2), 271–298.

McDermott, C. L., Acheampong, E., Arora-Jonsson, S., Asare, R., de Jong, W., Hirons, M., & Setyowati, A. (2019). SDG 16: Peace, justice and strong institutions-A political ecology perspective. *Sustainable development goals: Their impacts on forests and people* (pp. 510–540). Cambridge University Press.

Meadows, D. H., Meadows, D. L., Randers, J., & Behrens, W. W. (1972). *The limits to growth*. Universal Books.

Mensah, J. (2019). Sustainable development: Meaning, history, principles, pillars, and implications for human action: Literature review. *Cogent Social Sciences, 5*(1), 1653531.

Moldan, B., Janoušková, S., & Hák, T. (2012). How to understand and measure environmental sustainability: Indicators and targets. *Ecological Indicators, 17*, 4–13.

Monaco, E. (2023). *Global trends compendium: An essential guide to socio-economic and environmental change*. Springer Nature.

Morris, S. R. (2023). Teen mothers forgotten: The gap between high school and higher education. *Journal of Higher Education Policy and Leadership Studies, 4*(2), 107.

Moustairas, I., Vardopoulos, I., Kavouras, S., Salvati, L., & Zorpas, A. A. (2022). Exploring factors that affect public acceptance of establishing an urban environmental education and recycling center. *Sustainable Chemistry and Pharmacy, 25*, Article 100605.

Movilla-Pateiro, L., Mahou-Lago, X. M., Doval, M. I., & Simal-Gandara, J. (2021). Toward a sustainable metric and indicators for the goal of sustainability in agricultural and food production. *Critical Reviews in Food Science and Nutrition, 61*(7), 1108–1129.

Mumford, E. L., Martinez, D. J., Tyance-Hassell, K., Cook, A., Hansen, G. R., Labonté, R., & Parrish-Sprowl, J. (2023). Evolution and expansion of the One Health approach to promote sustainable and resilient health and well-being: A call to action. *Frontiers in Public Health, 10*, 1056459.

Panagiotopoulos, P., Vardopoulos, I., Maialetti, M., Ciaschini, C., Koundouri, P., & Salvati, L. (2024). Reimagining sustainable development and economic performance indicators: A human-centric Maslow-Bossel blueprint. *Economies, 12*(12), 338.

Paun, C. V., Musetescu, R. C., Topan, V. M., & Danuletiu, D. C. (2019). The impact of financial sector development and sophistication on sustainable economic growth. *Sustainability, 11*(6), 1713.

Perales, F., & Bernard, A. (2023). Continuity or change? How the onset of COVID-19 affected internal migration in Australia. *Population, Space and Place, 29*(2), Article e2626.

Perrin, F. (2022). Can the historical gender gap index deepen our understanding of economic development? *Journal of Demographic Economics, 88*(3), 379–417.

Pitt, S. J., & Gunn, A. (2024). The one health concept. *British Journal of Biomedical Science, 81*, 12366.

Pothen, F., & Welsch, H. (2019). Economic development and material use. Evidence from international panel data. *World Development, 115*, 107–119.

Quaranta, G., Salvia, R., Salvati, L., Paola, V. D., Coluzzi, R., Imbrenda, V., & Simoniello, T. (2020). Long-term impacts of grazing management on land degradation in a rural community of Southern Italy: Depopulation matters. *Land Degradation and Development, 31*(16), 2379–2394.

Ramankutty, N. (2018). Trends in global agricultural land use: Implications for environmental health and food security. *Annual Review of Plant Biology, 69*, 789–815.

Rao, K. R., & Rao, K. R. (2019). Geo-political Factors. *Wind energy for power generation: Meeting the challenge of practical implementation* (pp. 829–1071). Springer International Publishing.

Raworth, K. (2017). *Doughnut economics: Seven ways to think like a 21st-century economist*. Chelsea Green Publishing.

Redclift, M. (2005). Sustainable development (1987–2005): An oxymoron comes of age. *Sustainable Development, 13*(4), 212–227.

Ribeiro Hoffmann, A., Sandrin, P., & Doukas, Y. E. (Eds.). (2024). Climate change in regional perspective. European Union and Latin American initiatives, challenges, and solutions. Cham, Switzerland: United Nations University Series on Regionalism, Vol. 27, Springer Nature

Ruggerio, C. A. (2021). Sustainability and sustainable development: A review of principles and definitions. *Science of the Total Environment, 786*, Article 147481.

Sachs, J. D. (2015). *The age of sustainable development*. Columbia University Press.

Sallustio, L., Pettenella, D., Merlini, P., Romano, R., Salvati, L., Marchetti, M., & Corona, P. (2018). Assessing the economic marginality of agricultural lands in Italy to support land use planning. *Land Use Policy, 76*, 526–534.

Salvati, L., & Zitti, M. (2009). Substitutability and weighting of ecological and economic indicators: Exploring the importance of various components of a synthetic index. *Ecological Economics, 68*(4), 1093–1099.

Samhouri, D., Mahrous, H., Saidouni, A., El Kholy, A., Ghazy, R. M., Sadek, M., & Malkawi, M. (2025). Review on progress, challenges, and recommendations for implementing the One Health approach in the Eastern Mediterranean Region. *One Health*, 101057.

Sartorius, C. (2006). Second-order sustainability—conditions for the development of sustainable innovations in a dynamic environment. *Ecological Economics, 58*(2), 268–286.

Sathaye, J., Najam, A., Cocklin, C., Heller, T., Lecocq, F., Llanes-Regueiro, J., & Winkler, H. (2007). Sustainable development and mitigation. *Climate change 2007: Mitigation of climate change* (pp. 691–743). Cambridge University Press.

Silander, D. (2020). Agenda 2030 and the EU on peace, justice and strong institutions. *Implementing sustainable development goals in Europe* (pp. 162–184). Edward Elgar Publishing.

Singh, A., & Singh, V. (2024). The role of institutions in peace and justice for achieving sustainable development goal 16 and societal sustainability. *Amazonia Investiga, 13*(80), 9–18.

Singh, S. (1998). Adolescent childbearing in developing countries: A global review. *Studies in Family Planning, 29*(2), 117–136.

Šlaus, I., & Jacobs, G. (2011). Human capital and sustainability. *Sustainability, 3*(1), 97–154.

Swamy, L., Drazen, E., Johnson, W. R., & Bukoski, J. J. (2018). The future of tropical forests under the United Nations Sustainable Development Goals. *Journal of Sustainable Forestry, 37*(2), 221–256.

Tambo, E., Tang, S., Ai, L., & Zhou, X. N. (2017). The value of China-Africa health development initiatives in strengthening "One Health" strategy. *Global Health Journal, 1*(1), 33–46.

Tanzil, D., & Beloff, B. R. (2006). Assessing impacts: Overview on sustainability indicators and metrics. *Environmental Quality Management, 15*(4).

Teshome, M. (2024). The human health burden of climate change: Non-economic losses and ethical considerations towards achieving planetary health. *The Journal of Climate Change and Health, 19*, Article 100336.

Tomao, A., Quatrini, V., Corona, P., Ferrara, A., Lafortezza, R., & Salvati, L. (2017). Resilient landscapes in Mediterranean urban areas: Understanding factors influencing forest trends. *Environmental Research, 156*, 1–9.

Income Growth

2

Introduction

Gross Domestic Product (GDP) is the sum of final goods and services net of intermediate goods produced in a given economy over a fixed period (Kira, 2013). Its analysis primarily relies on the evolution of real GDP at constant prices, allowing for meaningful country-level comparisons over time (Fioramonti, 2013). From the demand side, GDP includes consumption, investment, public expenditure and net exports (Coyle, 2015). From the supply side, GDP is the sum of all the added values at each stage of production (Mazumdar, 2003). Per-capita GDP, i.e. the level of per-capita wealth produced within a territory over a given period, is obtained by dividing real GDP by the total population of the respective country or region (Balestra et al., 2018). This measure enables cross-country comparisons regardless of any difference in population size (Boarini & D'Ercole, 2013). In addition, it facilitates a comparative assessment of population living standards over time and across regions (e.g. Osberg & Sharpe, 2005). Economic growth, measured by GDP, is affected by multiple factors, e.g. capital accumulation, labour force expansion, technological advancements, and institutional frameworks (Doukas et al., 2025).

Among others, individual wealth is a crucial determinant of economic growth in all regions and countries (Dias & Tebaldi, 2012). Past experiences demonstrate that populations in high-income countries enjoy higher levels of health and longer life expectancy, as a possible result of more effective health policies. Within this context, recent studies highlight a positive correlation between health conditions and a country's ability to generate income (Hanefeld et al., 2018). Improved healthcare and better overall health outcomes contribute to increased productivity, labour market participation and higher per-capita GDP (Basel et al., 2020). Consequently, beyond the humanitarian imperatives of public health investments, these policies hold significant economic relevance, acting as catalysts for sustained income expansion. Besides health, education is another crucial factor in economic growth (Oketch, 2006).

Human capital theory suggests that investment in education leads to a more skilled and productive workforce, ultimately fostering innovation and economic expansion (Hess, 2016). Highly educated countries are likely to experience faster GDP growth due to increased labour efficiency and adaptability to technological progress. Furthermore, infrastructural development, referred to as transportation, energy and digital networks, plays a crucial role in enhancing economic productivity by reducing transaction costs and improving market efficiency (Hu et al., 2023). Sound financial systems and access to credit also support economic growth by facilitating entrepreneurship, investment, and resource allocation. Despite its widespread use as an economic indicator, GDP has notable limitations

C. Ciaschini and L. Salvati, *Sustainable Development*, Sustainable Development Goals Series,
https://doi.org/10.1007/978-3-032-12559-0_2

in assessing overall growth levels (Wang et al., 2022).

However, economic development extends beyond mere output growth, incorporating social progress (Ciaschini & Imbrenda, 2025), income distribution (Salvati, 2025), environmental sustainability (Sateriano et al., 2024) and institutional quality (Gupta et al., 2015). GDP, as an individual measure of economic growth, may fail to capture these multidimensional aspects, thus requiring further complementary indicators (Mujtaba, 2024). The United Nations Sustainable Development Goals (SDGs) emphasize a holistic approach to economic progress, particularly through SDG 8 (Kavouras et al., 2022). Within this context, the reliance on GDP as a primary metric seems to hinder the design of policies favouring the balance between economic expansion and socio-environmental sustainability (Ivković, 2016). A critical discussion in economic literature questions the persistence of sustainable long-term economic growth, echoing earlier concerns raised by Meadows et al. (1972), Redclift (2005), and Raworth (2017). The conventional economic model assumes continuous growth, often overlooking ecological constraints (Tsangaris et al., 2024). GDP-based growth models decouple economic expansion from environmental degradation (Ward et al., 2016). Therefore, achieving sustainability often requires slower growth and reduced consumption rates (Mishra et al., 2024).

Negative externalities such as pollution (Vinci et al., 2023), resource depletion (Salvati et al., 2025), and climate change (Ribeiro Hoffmann et al., 2024), challenge the notion of perpetual growth while fostering the need for alternative economic paradigms (Parrique et al., 2019). Green growth and circular economy models (Papamichael et al., 2023) have emerged as potential solutions, advocating economic systems that minimize waste (Vardopoulos, 2025), enhance resource efficiency (Sharma et al., 2021), and prioritize renewable energy sources (Vardopoulos et al., 2023). Another significant challenge in GDP measurement lies in addressing income inequality (Mukherjee, 2023). While per-capita GDP provides an average indication of a country's wealth, it does not account for disparities in income distribution (Charles et al., 2022). The Gini coefficient and other inequality indices should complement GDP by providing insights into wealth concentration and economic disparities within societies (Luptáčik & Nežinský, 2020). Rising inequality can hinder economic growth by reducing aggregate demand and preventing access to education and healthcare (Castells-Quintana et al., 2019).

Policymakers should design inclusive policies ensuring wealth distribution in line with broader development goals (Petrakos et al., 2024). Moreover, globalization significantly affected GDP growth patterns across countries (Doukas et al., 2024). Global financial markets, have played critical roles in shaping national economies (Wang et al., 2023). While globalization has led to economic expansion in many regions, it has also introduced vulnerabilities such as financial crises, trade imbalances and economic dependency on external markets (Abaidoo & Agyapong, 2022). The 2008 global financial crisis demonstrated how interconnected economies are susceptible to systemic shocks, necessitating stronger governance and regulatory frameworks, e.g. plans for economic resilience focussed on localized production and diversification strategies (Doukas et al., 2023; Goniewicz et al., 2023; Le & Hoang, 2022). Finally, technological advancements and digitalization are shaping economic structures and GDP dynamics (Abid, 2017). The development of artificial intelligence, automation, and digital platforms have improved productivity in several sectors while displacing traditional jobs (Vermeulen et al., 2018). Fin-tech innovations have introduced new growth opportunities (Aly, 2022), while raising concerns that regard labour market shifts, data privacy and digital divides among countries (Vardopoulos et al., 2024). Future policies should balance technological progress with labour market adaptability, ensuring an effective contribution of innovation to inclusive and sustainable economic growth (Clifton et al., 2020).

A comprehensive analysis of income expansion based on per-capita GDP growth rates over 60 years, from 1961 to 2020, was presented here. The original dataset, derived from the World Bank repository, initially comprised 265 statistical domains representing various countries and (autonomous government) territorial authorities. However, due to country aggregations and missing data hindering comparative analysis, the final sample of this study included 91 entities. The research employs univariate statistics and graphical analysis to examine fundamental trends in per-capita GDP over time. Additionally, bivariate analysis compares the distribution of GDP growth rates across countries in the starting year of each decade. Multivariate analysis (e.g. D'Agata et al., 2023; Karytsas et al., 2019; Moustairas et al., 2022; Vardopoulos, 2023), including Principal Component Analysis (PCA), finally aims at identifying patterns and underlying factors behind GDP growth.

Literature Review

Globally speaking, improvements in health lead to accelerated economic growth (Ranis et al., 2000). Nevertheless, such improvements could lead to modest increases in per-capita GDP, usually experienced in sufficiently long periods. Actions to contain some specific and serious diseases in developing countries do not relevantly affect this evidence (Labonté et al., 2010). Empirical works on the health impact on economic outcomes rely on data at both micro (e.g. household) and macro (e.g. country) levels. From a microeconomic perspective, the individual's health is crucial for economic performance. At the macroeconomic level, improved healthcare systems have usually been associated with affluent countries (Wenjuan & Zhao, 2023). Nevertheless, a reliable macroeconomic conclusion could not only emerge from microeconomic or cross–sectional evidence, since microeconomic analysis considers the effect of changes in population health on general equilibrium.

Economic Growth and Healthcare Conditions

In general, health affects per-capita income in lagged periods, thus explaining inequalities in healthcare services distribution between countries (Permanyer & Scholl, 2019). Within this context, an inverse relationship has been experienced in the 1950–1990 time interval (i.e. until the advent of AIDS epidemic), between inequality in healthcare distribution and income inequalities at a country level (Tabutin et al., 2017). The second quarter of 2020 registered a severe recession in the global economy because of Covid-19 (Riouch et al., 2024), then recovered in the third quarter, only to decline further in fall 2020. Besides Covid-19, different diseases also affected different countries, implementing heterogeneous policies to deal with them (Ibn-Mohammed et al., 2021). König and Winkler (2021) examined the extent to which growth in the first three quarters of 2020 reflects the varying intensity of government restrictions. Some countries preferred a lighter application of social distancing with the aim at reducing the costs connected to mandatory restrictions in a climate of reduced health risk (Gupta et al., 2023). From a purely economic perspective, hard lockdowns should be avoided as any rise in lockdown intensity negatively affects economic activity.

In a broader economic context (Yusuf & Nasrulddin, 2024), healthcare improvements also influence job market dynamics, workforce participation, and human capital accumulation. A healthier workforce is more productive and less prone to absenteeism reducing economic costs associated with illness-related work disruptions (Jakovljevic et al., 2021). Additionally, investments in healthcare infrastructure, medical research, and pharmaceutical advancements contribute to economic diversification by stimulating job creation and technological innovation (Hidayat et al., 2024). Countries with robust healthcare systems tend to exhibit higher economic resilience in times of crisis, mitigating the adverse effects of pandemics and other health shocks (Glonti et al., 2015). Moreover, disparities in healthcare access across regions contribute to income inequality, as populations with limited healthcare services often experience lower economic mobility and weaker job market outcomes (Audi, 2022). Policies aimed at enhancing universal healthcare access, and preventive medicine can thus function as long-term economic growth strategies (Baeten et al., 2018). Historical data suggests that countries prioritizing public health initiatives tend to experience higher GDP growth rates over time due to improved human capital formation and increased life expectancy, which fosters a more stable and productive workforce (Graff Zivin & Neidell, 2013).

Healthcare expenditures, however, pose significant fiscal challenges (Lopez-Claros, 2014). Governments should balance the economic benefits of healthcare investments with budgetary constraints and efficiency considerations (Raghupathi & Raghupathi, 2020). Rising healthcare costs, driven by aging populations and medical advancements, put pressure on public finances, requiring policymakers to implement sustainable funding mechanisms (Lee et al., 2021). The role of private sector involvement in healthcare, alongside public investments, remains a topic of debate, particularly regarding cost-effectiveness (Delardas et al., 2022; Mosca et al., 2016; Russo et al., 2021). Additionally, international cooperation in health policy plays a vital role in addressing global health challenges and mitigating economic risks associated with pandemics. Collaborative efforts in vaccine distribution, disease surveillance and health system strengthening enhance global economic stability by reducing the spread of infectious diseases (Bloom & Cadarette, 2020) and minimizing economic disruptions (Decouttere et al., 2021).

International financial institutions have emphasized the need for integrated health and economic strategies to ensure sustainable development (Modibbo & Inuwa, 2020). Ultimately, the intersection of health and economics underscores the necessity of a holistic approach to policymaking (Koon et al., 2016). While short-term economic growth considerations may discourage extensive public health expenditures (Gottlieb et al., 2020), the long-term benefits of a healthier population in terms of productivity, innovation and overall economic stability cannot be overlooked (Abdulm et al. 2024). Future research should investigate health improvements and economic dynamics more deeply (Ramani & Mavalankar, 2006), especially under evident demographic shifts, and emerging infectious diseases (Siddiqi et al., 2023).

Economic Growth and Environmental Degradation

In affluent countries, dealing with environmental issues at later stages of their development process may discourage a more environment-friendly approach (Ahmad et al., 2024; Clark et al., 2018; Sun et al., 2023). Under such perspective, the inverted-U shape relationship existing between income and environmental quality and better known as the Environmental Kuznets Curve (EKC), highlights that economic growth goes along with environmental degradation (Bosetti et al., 2011; Kahuthu, 2006; Kaika & Zervas, 2013; Sadik-Zada & Gatto, 2023). Development strategies have usually preferred economic growth to environmental policy (Eliades et al., 2022; Mignamissi et al., 2024; Soubbotina, 2004). The EKC model was frequently observed in the case of CO_2 emissions, properly considering environmental issues within their development policy (Leal & Marques, 2022; Meyfroidt & Lambin, 2011; Vardopoulos & Konstantinou, 2017).

A higher level of integration within the international market should correspond to a higher level of CO_2 emissions (Hoang & Kanemoto, 2021) and a faster process of deforestation. Globalization should combine economic development with environmental regulation (Stojkoski et al., 2023). A more sustainable global economy requires the integration of environmental principles in the management priorities (Egidi et al., 2020; Gallego-Álvarez et al., 2014; Sallustio et al., 2018). Furthermore, the role of technological advancements in shaping the relationship between economic growth and environmental sustainability should be considered (Liang & Yang, 2019). Innovations in waste management (Vardopoulos et al., 2025), and renewable energy and energy efficiency (Vardopoulos, 2018), contribute to breaking the traditional EKC patterns, enabling economic growth without the corresponding environmental degradation (Khan et al., 2020; Salvati & Zitti, 2009; Zambon & Salvati, 2019).

Countries actively investing in green technology and adopting strict environmental policies have shown signs of an earlier turning point in the EKC (Kpurunee et al., 2024), proving that proactive environmental management can decouple growth from pollution (Ogundipe et al., 2014). Moreover, financial markets and institutional policies play a crucial role in the relationship between economic development and environmental quality (Das & Sethi, 2023). Green bonds, carbon credits and impact investments have emerged as tools to promote sustainable economic activities while curbing environmental harm (Khan et al., 2024). The financial sector, therefore, has the potential to accelerate the transition towards an environmentally conscious global economy by incentivizing sustainable business models (Ghisellini et al., 2016). Policy perspectives, in turn emphasize the need for countries to align economic growth with ecological preservation (Ahmad et al., 2022). Implementing environmental projects are essential mechanisms promoting this transition (Wang et al., 2025). However, such measures largely depends on political actions, public awareness, and global cooperation (Overmars et al., 2014).

In conclusion, while the EKC model provides an initial framework describing the interplay between economic growth and environmental quality (e.g. Salvati & Carlucci, 2014), it is not a deterministic rule (Azam & Khan, 2016). The trajectory of environmental impact varies based on government policies, technological advancements, and global economic integration (Tamazian & Rao, 2010). Future research should focus on identifying strategies that enable sustainable development without relying on the traditional 'grow now, clean later' paradigm (Sadik-Zada & Gatto, 2023). A balanced approach fostering economic prosperity while ensuring environmental sustainability is crucial for long-term global well-being (Ossewaarde & Ossewaarde-Lowtoo, 2020).

Empirical Findings

Global economic dynamics highlight a discontinuous path of GDP growth rates over time. Since World War II, negative peaks were observed in 1975—when the oil crisis triggered increasing energy prices—and in 1982—when heavy recession caused a sharp decrease in GDP in most Western countries. Relevant economic crises with consequent decreases in GDP in many countries around the world also occurred in 1991 and 2009. The last shock was observed in 2020 because of Covid-19 pandemic. A refined analysis of such patterns was performed using statistical instruments, from descriptive statistics to multidimensional analysis, as described below.

Descriptive Statistics

Table 2.1 illustrates a selection of descriptive statistics for the first year of seven decades between 1961 and 2020. The average values fluctuate over time, without exhibiting any defined trend. The lowest value occurred in 1980 (−30%), while the highest value (43%) was observed in 2020. The highest value of median growth rates was observed in 2010 (3.5%); the minimum value (−4.5%) was instead recorded for 2020, fluctuating over the last decade. Negative values of skewness were observed for 1961, 1980, 1990, and 2000, shifting to positive figures in the remaining cases. Kurtosis assumed positive values, except for 1990.

Figure 2.1 represents a mean and whisker plot illustrating standard deviations in the statistical distribution of the investigated variable. This plot highlights a heterogeneous path in the global expansion of incomes over time. More specifically, the observations for both 1976 and 2010 stand out when compared with the rest of the time series. Economic expansion in 1976, after a major oil crisis, was particularly relevant in a global scale. In 2010, both affluent economies and emerging countries experienced a sustained recovery rate, after

Table 2.1 Descriptive statistics representing the distribution of per-cent growth rate in per-capita gross domestic product (GDP) across world countries

Metric	1961	1970	1980	1990	2000	2010	2020
Mean	1.8	4.3	1.5	0.8	1.9	3.6	− 4.9
Minimum	− 26.5	− 8.5	− 30.0	− 9.7	− 9.3	− 7.1	− 19.2
Maximum	13.7	22.2	14.3	10.7	11.1	18.1	42.8
Normalized range	22.3	7.2	30.2	25.1	10.8	6.9	− 12.6
Median	2.2	3.3	1.6	1.3	2.6	3.5	− 4.5
25 percentile	0.1	0.8	− 1.9	− 2.6	− 0.1	1.4	− 8.2
75 percentile	4.7	6.6	5.1	3.5	3.7	5.6	− 2.0
Skewness	− 1.7	0.8	− 1.6	− 0.2	− 0.4	0.5	3.7
Kurtosis	6.1	1.9	7.5	− 0.1	0.6	2.7	26.7

Source Authors' elaboration of World Bank indicators

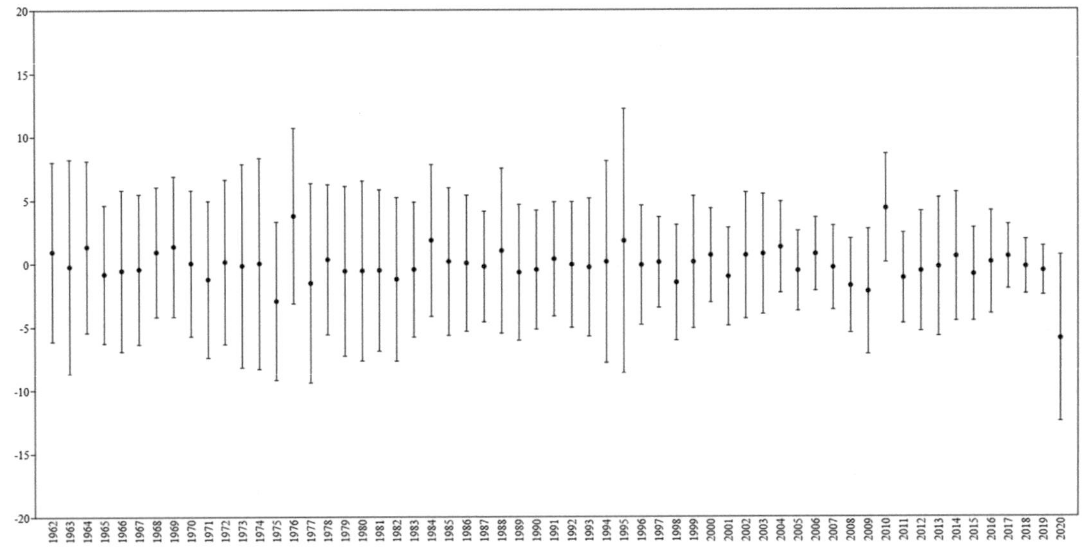

Fig. 2.1 A graphical summary of the statistical distribution of per-capita GDP growth rates across world countries based on a mean and whisker plot. *Source* Authors' elaboration of World Bank indicators

the worldwide spread of the United States crisis since 2006. On the contrary, particularly negative and spatially coherent values were observed for 2020, because of Covid-19 pandemic.

Correlation Analysis

Considering correlation trends, parametric (Bravais-Pearson linear) and non-parametric (Spearman rank) coefficients (Fig. 2.2) have displayed rather similar values along the investigated time series up to the early 1990s, and highlight the importance of linear development paths across world countries. Different behaviours have emerged between 1991 and 2012, with the dominance of nonlinear (negative) correlation coefficients, indicating a diverging development path across world economic systems. A particularly evident heterogeneity in GDP growth rates, with the dominance of mild (and statistically insignificant) coefficients, emerged between 2013 and 2020. Considering only the first and last observations of the time series, the average value of the studied indicator became negative in 2020, likely because of Covid-19 pandemic. By

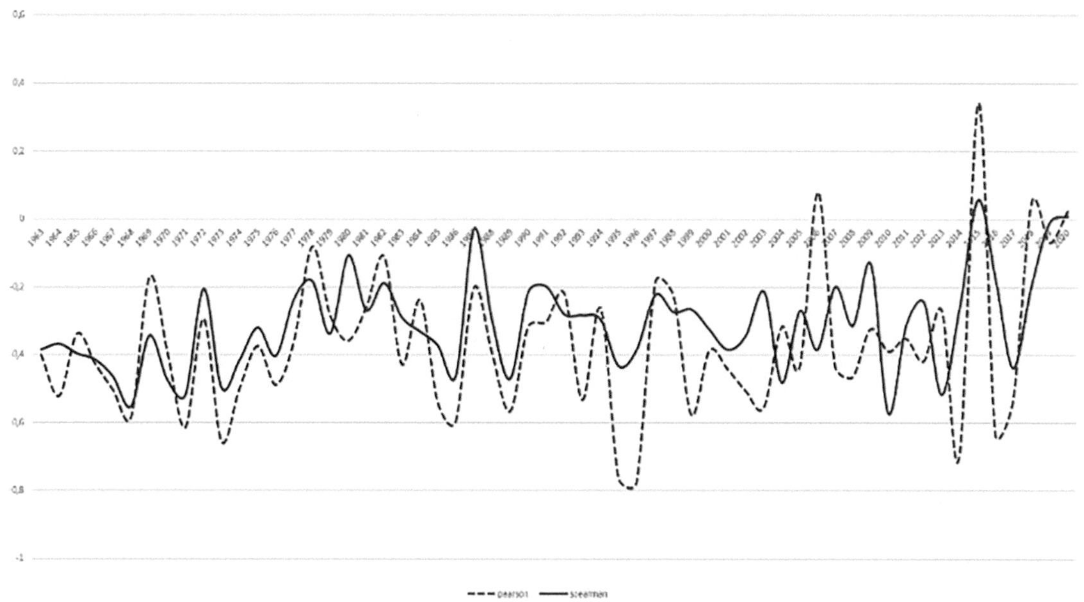

Fig. 2.2 Trends over time in parametric (Pearson) and nonparametric (Spearman) correlation coefficients within the (−1) lagged time series of per-capita GDP growth. *Source* Authors' elaboration of World Bank indicators

comparing the statistical distribution of incomes at these two years, countries experiencing stagnant economic dynamics have also displayed similar values of population growth (e.g. Benin, Haiti, Rwanda). Other countries, such as Myanmar, Turkey, Kenya, Congo, Burundi and Algeria, experienced an important increase in per-capita GDP growth rate in 2020 compared to 1961.

Multidimensional Analysis

Based on the results of a Principal Component Analysis (PCA) analysing GDP growth rates over time, the most relevant dimensions (and, especially, PC1 and PC2) were proved to be sufficiently informative when identifying and delineating the most critical drivers of GDP growth rates, while additional PCs displayed a lower explanatory value. PC1 performed the highest eigenvalue, capturing the largest amount of variability in the original data matrix. Eigenvalues gradually decreased across the selected PCs. PC3 explained 6.5% of the total variance; each of the remaining components contributed progressively with smaller proportions of explained variance. Based on this eigenvalue function, two components were selected: PC1, explaining 12.1% of the total variance, and PC2, contributing 8.7% to the total variance; taken together, these two components led to a cumulative proportion of explained variance amounting, approximately, to 21%. Considering the relatively high matrix dimension (years by column and observations (countries) by row), this result is taken as appreciable in a specific exploratory analysis perspective. Within this context, PC1 and PC2 may describe—likely better than other analysis' dimensions—the dominant macroeconomic factors, such as trade, investment, and inflation, affecting GDP growth rates and determining significant economic disparities across countries.

Figure 2.3 shows a biplot of the first two components, highlighting the geographical pattern of GDP growth across countries and suggesting an extensive interpretation of its determinants. A dense cluster of observations near the plot's origin indicates countries and years with average GDP growth rates, which are rather coherent with the global mean reflective of the 'normal' rate of economic expansion observed globally after World War II. Outliers' countries, as China (CHN),

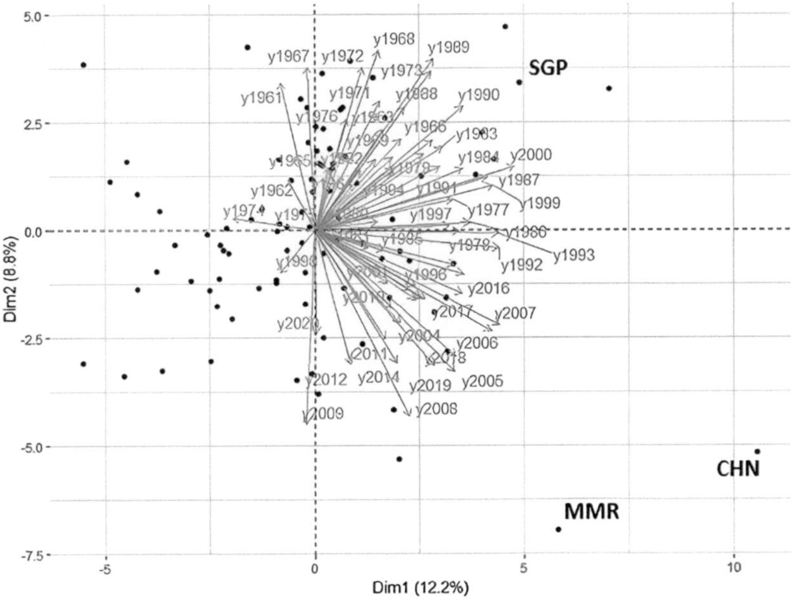

Fig. 2.3 Results of a Principal Component Analysis as illustrated in a biplot reflecting the distribution of variables' (i.e. years) loadings and rows (i.e. countries) scores along the first two axes. *Source* Authors' elaboration of World Bank indicators

Myanmar (MMR) and Singapore (SGP), located rather far from the origin, showed peculiar characteristics related to years of exceptionally high (or, by contrast, restricted) GDP growth. These countries experienced significant events like economic crises, post-stagnation rebounds, or large-scale socioeconomic reforms (e.g. financial recessions or structural adjustments in emerging countries). The grey arrows represent variables more likely tied to GDP growth (possibly including investment, trade, inflation, and population growth dynamics). Longer arrows describe variables providing the strongest contributions to principal components, having shown a significant (explanatory) power for GDP growth trends. Observations in line with key (positive) growth factors (e.g. trade openness, investment) were mostly observed for individual years or countries with robust GDP growth, as some emerging Asian economies in recent decades.

Discussion

The present work depicts the heterogeneous nature of economic expansion, both at the global and regional levels. Since World War II, global per-capita GDP has generally performed a strong upward trajectory, interrupted only by major global crises of economic, financial, or energy-related origins (Banerji, 2013; Brown, 2014; Koser et al., 2018). Sustained growth has led to widespread improvements in living standards, despite a continuously expanding global population (Clark, 2011). However, income distribution remains highly uneven, with pronounced social inequalities emerging as a consequence of unbalanced economic growth (e.g. Coluzzi et al., 2022; Nickayin et al., 2022; Quaranta et al., 2020). In some emerging countries, populations continue to face relevant challenges, as malnutrition, limited access to education, and higher mortality rates, exacerbating socioeconomic disparities (Nguy˜ên & Phan, 2025).

The most recent and striking example of global economic disruption occurred in 2020. In general, pandemics led to severe recessions in many economies, with some nations experiencing prolonged contractions due to inadequate healthcare infrastructure and prolonged lockdowns. Several countries, such as Argentina and Greece, have been consistently affected by economic instability, recording negative GDP growth rates due to persistent financial crises and structural inefficiencies. In Africa, economic growth has been highly variable across the 54 analysed states. While conflict-ridden nations have faced economic stagnation, other countries have emerged as some of the world's fastest-growing economies, demonstrating potential for significant future development (Ali & Mohsin, 2023; Ji et al., 2022; Lyakurwa, 2009). Despite slow progress (Lee et al., 2023), many African nations are prone to becoming dynamic emerging markets with substantial growth potential decline (Ahmad et al., 2025).

While GDP growth remains a key indicator for assessing economic trends and national performance, it is increasingly recognized that sustainable development requires broader measures beyond GDP (Fioramonti, 2016). A comprehensive evaluation of regional and national progress should integrate indicators related to environmental sustainability and social well-being (Zhan & Santos-Paulino, 2021). With this perspective in mind, the United Nations Sustainable Development Goals (SDGs) framework provides a holistic approach (Fenner & Cernev, 2021). A more detailed analysis of economic expansion also reveals the need to consider external shocks and structural adjustments. Economic crises, trade liberalization policies, and shifts in global supply chains significantly impact national GDP growth patterns (Spence, 2024). Additionally, regional disparities highlight the importance of tailored policy interventions (Tödtling & Trippl, 2005). For instance, economies heavily reliant on commodity exports may experience volatile growth due to fluctuations in global commodity prices (Manta et al., 2024), while diversified economies tend to exhibit more stable GDP trends (Giordono et al., 2019). The empirical results of Principal Component Analysis have also demonstrated how macroeconomic fundamentals, such as inflation control, fiscal stability, and investment in human capital, play crucial roles in determining long-term growth trajectories (Rodrigo-Comino et al., 2021). Countries successfully handling financial crises and implementing effective economic policies tend to recover faster and maintain sustainable growth (e.g. Salvati et al., 2016). In contrast, economies suffering from prolonged political instability, governance challenges (Hariram et al., 2023), or structural inefficiencies may struggle to sustain long-term growth, even with favourable global conditions (Nguyen, 2024).

Concluding Remarks

The present study underscores the complexity of economic growth and its dependence on various interrelated factors (Zhang et al., 2024). The findings advocate for a multidimensional approach in assessing economic performance (Bajaj & Bhooshetty, 2024), incorporating both traditional GDP-based measures and alternative indicators accounting for broader social, environmental, and economic dimensions (Spilioti & Anastasiou, 2024). Given the growing awareness of sustainability concerns (Jackson, 2016), future economic policies should stimulate environmental stewardship (Aiguobarueghian et al., 2024). Policymakers should prioritize strategies that enhance productivity while promoting responsible resource management and equitable wealth distribution (Rahman & Alam, 2021). The challenge ahead lies in fostering economic expansion that is both resilient and inclusive (Kohli & Agarwala, 2017), ensuring that prosperity reaches all segments of society while safeguarding the planet for future generations.

References

Abaidoo, R., & Agyapong, E. K. (2022). Financial development and institutional quality among emerging economies. *Journal of Economics and Development, 24*(3), 198–216.

Abdul, S., Adeghe, E. P., Adegoke, B. O., Adegoke, A. A., & Udedeh, E. H. (2024). Promoting health and educational equity: Cross-disciplinary strategies for enhancing public health and educational outcomes. *World Journal of Biology Pharmacy and Health Sciences, 18*(2), 416–433.

Abid, M. (2017). Does economic, financial and institutional developments matter for environmental quality? A comparative analysis of EU and MEA countries. *Journal of Environmental Management, 188*, 183–194.

Ahmad, M., Ahmed, Z., & Luo, C. (2024). Natural resources, economic globalization, and sustainable development: Can economic complexity and environmental regulations cure the resource curse? In: *Natural*

Resources Forum (Vol. 48, No. 4, pp. 1417–1435). Oxford, UK: Blackwell.

Ahmad, A. U., Jeevan, J., Ruslan, S. M. M., & Abubakar, A. M. (2025). The asymmetric nexus of energy factors, green economy factors, blue economy factors, and african growth sustainability: An empirical evidence from hidden cointegration approach. *Journal of the Knowledge Economy*, 1–49.

Ahmad, M., Ahmed, Z., Yang, X., Hussain, N., & Sinha, A. (2022). Financial development and environmental degradation: Do human capital and institutional quality make a difference? *Gondwana Research, 105*, 299–310.

Aiguobarueghian, I., Adanma, U. M., Ogunbiyi, E. O., & Solomon, N. O. (2024). An overview of initiatives and best practices in resource management and sustainability. *World Journal of Advanced Research and Reviews, 22*(2), 1734–1745.

Ali, S. B., & Mohsin, A. (2023). Exploring financial soundness and economic growth dynamics in Pakistan. *Journal of Business and Economic Options, 6*(1), 1–15.

Aly, H. (2022). Digital transformation, development and productivity in developing countries: Is artificial intelligence a curse or a blessing? *Review of Economics and Political Science, 7*(4), 238–256.

Audi, M. (2022). Income inequality and health across the Mediterranean: A panel data analysis. *Journal of Business and Economic Options, 5*(2), 23–28.

Azam, M., & Khan, A. Q. (2016). Testing the environmental Kuznets curve hypothesis: A comparative empirical study for low, lower middle, upper middle and high income countries. *Renewable and Sustainable Energy Reviews, 63*, 556–567.

Baeten, R., Spasova, S., Vanhercke, B., & Coster, S. (2018). Inequalities in access to healthcare. *A study of national policies, European Social Policy Network (ESPN).* Brussels: European Commission.

Bajaj, A., & Bhooshetty, L. (2024). Unleashing economic potential: Decoding the FDI-economic growth nexus in G-15 economies amidst unique host country traits. *Asia-Pacific Journal of Regional Science, 8*(2), 651–679.

Balestra, C., Boarini, R., & Ruiz, N. (2018). Going beyond GDP: Empirical findings. *Handbook of research on economic and social well-being* (pp. 52–103). Edward Elgar.

Banerji, H. (2013). Globalization can lead to inclusive growth in: Developing countries-myth or reality. *Economic and Social Development: Book of Proceedings*, 43.

Basel, S., Gopakumar, K. U., & Rao, R. P. (2020). Broad-based index for measurement of development. *Journal of Social and Economic Development, 22*, 182–206.

Bloom, D. E., & Cadarette, D. (2020). Strengthening the global response to infectious disease threats in the twenty-first century, with a COVID-19 epilogue. *Global Health*, 51–75.

Boarini, R., & D'Ercole, M. M. (2013). Going beyond GDP: An OECD perspective. *Fiscal Studies, 34*(3), 289–314.

Bosetti, V., Lubowski, R., Golub, A., & Markandya, A. (2011). Linking reduced deforestation and a global carbon market: Implications for clean energy technology and policy flexibility. *Environment and Development Economics, 16*(4), 479–505.

Brown, M. M. (2014). Building partnerships for an inclusive globalization. *The Future of Globalization* (pp. 58–61). Routledge.

Castells-Quintana, D., Royuela, V., & Thiel, F. (2019). Inequality and sustainable development: Insights from an analysis of the human development index. *Sustainable Development, 27*(3), 448–460.

Charles, V., Gherman, T., & Paliza, J. C. (2022). The Gini Index: A modern measure of inequality. *Modern indices for international economic diplomacy* (pp. 55–84). Springer International Publishing.

Ciaschini, C., & Imbrenda, V. (2025). *Local development and socio-environmental systems. Experiences and practices toward sustainability in Mediterranean Europe.* River Publishers.

Clark, R. (2011). World health inequality: Convergence, divergence, and development. *Social Science and Medicine, 72*(4), 617–624.

Clark, R., Reed, J., & Sunderland, T. (2018). Bridging funding gaps for climate and sustainable development: Pitfalls, progress and potential of private finance. *Land Use Policy, 71*, 335–346.

Clifton, J., Glasmeier, A., & Gray, M. (2020). When machines think for us: The consequences for work and place. *Cambridge Journal of Regions, Economy and Society, 13*(1), 3–23.

Coluzzi, R., Bianchini, L., Egidi, G., Cudlin, P., Imbrenda, V., Salvati, L., & Lanfredi, M. (2022). Density matters? Settlement expansion and land degradation in Peri-urban and rural districts of Italy. *Environmental Impact Assessment Review, 92*, Article 106703.

Coscieme, L., Mortensen, L. F., Anderson, S., Ward, J., Donohue, I., & Sutton, P. C. (2020). Going beyond gross domestic product as an indicator to bring coherence to the sustainable development goals. *Journal of Cleaner Production, 248*, Article 119232.

Coyle, D. (2015). *GDP: A brief but affectionate history-revised and* (expanded). Princeton University Press.

D'Agata, A., Ponza, D., Strolu, F. A., Vardopoulos, I., Rontos, K., Escrivà, F., Chelli, F., Alaimo, L. S., Luca, S., & Nickayin, S. S. (2023). Toward sustainable development trajectories? Estimating urban footprints from high-resolution Copernicus layers in Athens Greece. *Land, 12*(8), 1490.

Das, A., & Sethi, N. (2023). Modelling the environmental pollution-institutional quality nexus in low- and middle-income countries: Exploring the role of financial development and educational level. *Environment, Development and Sustainability, 25*(2), 1492–1518.

Decouttere, C., De Boeck, K., & Vandaele, N. (2021). Advancing sustainable development goals through immunization: A literature review. *Globalization and Health, 17*(1), 95.

Delardas, O., Kechagias, K. S., Pontikos, P. N., & Giannos, P. (2022). Socio-economic impacts and challenges of the coronavirus pandemic (COVID-19): An updated review. *Sustainability, 14*(15), 9699.

Dias, J., & Tebaldi, E. (2012). Institutions, human capital, and growth: The institutional mechanism. *Structural Change and Economic Dynamics, 23*(3), 300–312.

Doukas, Y. E., Maravegias, N., Ioannides, S., Salvati, L., & Vardopoulos, I. (2025). Economic resilience and sustainability in the European Union: A comparative analysis of pre-and post-economic crisis. *Environmental Sustainability and Global Change* (pp. 257–273). Elsevier.

Doukas, Y. E., Maravegias, N., Salvati, L., & Vardopoulos, I. (2024). Resilienza economica e sostenibilità: Indicatori macroeconomici e dinamiche politiche pre-e post-crisi nell'Unione Europea. *Documenti Geografici, 3*, 491–525.

Doukas, Y. E., Salvati, L., & Vardopoulos, I. (2023). Unraveling the European agricultural policy sustainable development trajectory. *Land, 12*(9), 1749.

Egidi, G., Salvati, L., & Vinci, S. (2020). The long way to tipperary: City size and worldwide urban population trends, 1950–2030. *Sustainable Cities and Society, 60*, Article 102148.

Eliades, F., Doula, M. K., Papamichael, I., Vardopoulos, I., Voukkali, I., & Zorpas, A. A. (2022). Carving out a niche in the sustainability confluence for environmental education centers in Cyprus and Greece. *Sustainability, 14*(14), 8368.

Fenner, R., & Cernev, T. (2021). The implications of the Covid-19 pandemic for delivering the sustainable development goals. *Futures, 128*, Article 102726.

Fioramonti, L. (2013). *Gross domestic problem: The politics behind the world's most powerful number*. Bloomsbury.

Fioramonti, L. (2016). A post-GDP world? Rethinking international politics in the 21st century. *Global Policy, 7*(1), 15–24.

Gallego-Álvarez, I., Vicente-Galindo, M. P., Galindo-Villardón, M. P., & Rodríguez-Rosa, M. (2014). Environmental performance in countries worldwide: Determinant factors and multivariate analysis. *Sustainability, 6*(11), 7807–7832.

Ghisellini, P., Cialani, C., & Ulgiati, S. (2016). A review on circular economy: The expected transition to a balanced interplay of environmental and economic systems. *Journal of Cleaner Production, 114*, 11–32.

Giordono, L. S., Jones, M. D., & Rothwell, D. W. (2019). Social policy perspectives on economic inequality in wealthy countries. *Policy Studies Journal, 47*, S96–S118.

Glonti, K., Gordeev, V. S., Goryakin, Y., Reeves, A., Stuckler, D., McKee, M., & Roberts, B. (2015). A systematic review on health resilience to economic crises. *PLoS ONE, 10*(4), Article e0123117.

Goniewicz, K., Khorram-Manesh, A., Burkle, F. M., Hertelendy, A. J., & Goniewicz, M. (2023). The European Union's post-pandemic strategies for public health, economic recovery, and social resilience. *Global Transitions, 5*, 201–209.

Gottlieb, N., Trummer, U., Davidovitch, N., Krasnik, A., Juárez, S. P., Rostila, M., & Bozorgmehr, K. (2020). Economic arguments in migrant health policymaking: Proposing a research agenda. *Globalization and Health, 16*, 1–5.

Graff Zivin, J., & Neidell, M. (2013). Environment, health, and human capital. *Journal of Economic Literature, 51*(3), 689–730.

Gupta, J., Pouw, N. R., & Ros-Tonen, M. A. (2015). Towards an elaborated theory of inclusive development. *The European Journal of Development Research, 27*, 541–559.

Gupta, S., Montenovo, L., Nguyen, T., Lozano-Rojas, F., Schmutte, I., Simon, K., & Wing, C. (2023). Effects of social distancing policy on labor market outcomes. *Contemporary Economic Policy, 41*(1), 166–193.

Hanefeld, J., Mayhew, S., Legido-Quigley, H., Martineau, F., Karanikolos, M., Blanchet, K., & Balabanova, D. (2018). Towards an understanding of resilience: Responding to health systems shocks. *Health Policy and Planning, 33*(3), 355–367.

Hariram, N. P., Mekha, K. B., Suganthan, V., & Sudhakar, K. (2023). Sustainalism: An integrated socio-economic-environmental model to address sustainable development and sustainability. *Sustainability, 15*(13), 10682.

Hess, P. N. (2016). *Economic growth and sustainable development*. Routledge.

Hidayat, B. A., Supartoyo, Y. H., Setiawan, S., Ragimun, R., & Salim, Z. (2024). Government infrastructure investment stimulation through booming natural resources: Evidence from a lower-middle-income country. *PLoS ONE, 19*(5), Article e0301710.

Hoang, N. T., & Kanemoto, K. (2021). Mapping the deforestation footprint of nations reveals growing threat to tropical forests. *Nature Ecology and Evolution, 5*(6), 845–853.

Hu, J., Zhang, H., & Irfan, M. (2023). How does digital infrastructure construction affect low-carbon development? A multidimensional interpretation of evidence from China. *Journal of Cleaner Production, 396*, Article 136467.

Ibn-Mohammed, T., Mustapha, K. B., Godsell, J., Adamu, Z., Babatunde, K. A., Akintade, D. D., & Koh, S. C. L. (2021). A critical analysis of the impacts of COVID-19 on the global economy and ecosystems and opportunities for circular economy strategies. *Resources, Conservation and Recycling, 164*, Article 105169.

Ivković, A. F. (2016). Limitations of the GDP as a measure of progress and well-being. *Ekonomski Vjesnik/econviews–review of Contemporary Business, Entrepreneurship and Economic Issues, 29*(1), 257–272.

Jackson, T. (2016). *Prosperity without growth: Foundations for the economy of tomorrow*. Routledge.

Jakovljevic, M., Wu, W., Merrick, J., Cerda, A., Varjacic, M., & Sugahara, T. (2021). Asian innovation in pharmaceutical and medical device industry–beyond tomorrow. *Journal of Medical Economics, 24*(1), 42–50.

Ji, X., Dong, F., Zheng, C., & Bu, N. (2022). The influences of international trade on sustainable economic growth: An economic policy perspective. *Sustainability, 14*(5), 2781.

Kahuthu, A. (2006). Economic growth and environmental degradation in a global context. *Environment, Development and Sustainability, 8*(1), 55–68.

Kaika, D., & Zervas, E. (2013). The Environmental Kuznets Curve (EKC) theory—Part A: Concept, causes and the CO_2 emissions case. *Energy Policy, 62*, 1392–1402.

Karytsas, S., Vardopoulos, I., & Theodoropoulou, E. (2019). Factors affecting sustainable market acceptance of residential microgeneration technologies. A two time period comparative analysis. *Energies, 12*(17), 3298.

Kavouras, S., Vardopoulos, I., Mitoula, R., Zorpas, A. A., & Kaldis, P. (2022). Occupational health and safety scope significance in achieving sustainability. *Sustainability, 14*(4), 2424.

Khan, M. A., Kong, D., Xiang, J., & Zhang, J. (2020). Impact of institutional quality on financial development: Cross-country evidence based on emerging and growth-leading economies. *Emerging Markets Finance and Trade, 56*(15), 3829–3845.

Khan, M. A., Máté, D., Abdulahi, M. E., Sadaf, R., Khan, M. A., Popp, J., & Oláh, J. (2024). Do institutional quality, innovation and technologies promote financial market development? *European Journal of International Management, 22*(3), 484–507.

Kira, A. R. (2013). The factors affecting Gross Domestic Product (GDP) in developing countries: The case of Tanzania. *European Journal of Business and Management, 5*(4), 148–158.

Kohli, H. S., & Agarwala, R. (2017). *The world in 2050: Striving for a more just, prosperous, and harmonious global community*. Oxford University Press.

König, M., & Winkler, A. (2021). COVID-19: Lockdowns, fatality rates and GDP growth. *Intereconomics, 56*(1), 32–39.

Koon, A. D., Hawkins, B., & Mayhew, S. H. (2016). Framing and the health policy process: A scoping review. *Health Policy and Planning, 31*(6), 801–816.

Koser, R., Sarwar, S., & Siddiqi, W. (2018). Economics of globalization: An approach towards inclusive growth for developing nations. *Pakistan Journal of Applied Economics, 28*(1), 255–267.

Kpurunee, G. L., Nwikiabeh, L. B., & Osho, A. (2024). Green policies and citizenship: The route to national development. *Nigerian Academy of Management Journal, 19*(1), 53–61.

Labonté, R., & Gagnon, M. L. (2010). Framing health and foreign policy: Lessons for global health diplomacy. *Globalization and Health, 6*, 1–19.

Le, H. T., & Hoang, D. P. (2022). Economic sanctions and environmental performance: The moderating roles of financial market development and institutional quality. *Environmental Science and Pollution Research, 29*(13), 19657–19678.

Leal, P. H., & Marques, A. C. (2022). The evolution of the environmental Kuznets curve hypothesis assessment: A literature review under a critical analysis perspective. *Heliyon, 8*(11), Article e11521.

Lee, B., Kim, H., & Tavakoli, A. (2023). The impact of economic growth, inflation and unemployment on subjective financial satisfaction: A New global evidence. *Cogent Economics and Finance, 11*(2), 2287908.

Lee, D. C., Liang, H., & Shi, L. (2021). The convergence of racial and income disparities in health insurance coverage in the United States. *International Journal for Equity in Health, 20*, 1–8.

Liang, W., & Yang, M. (2019). Urbanization, economic growth and environmental pollution: Evidence from China. *Sustainable Computing: Informatics and Systems, 21*, 1–9.

Lopez-Claros, A. (2014). Fiscal challenges after the global financial crisis: A survey of key issues. *Journal of International Commerce, Economics and Policy, 5*(02), 1450004.

Luptáčik, M., & Nežinský, E. (2020). Measuring income inequalities beyond the Gini coefficient. *Central European Journal of Operations Research, 28*(2), 561–578.

Lyakurwa, W. (2009). Prospects for economic governance: Resilient pro-poor growth. *Foresight, 11*(4), 66–81.

Manta, A. G., Doran, N. M., Bădîrcea, R. M., Badareu, G., Gherțescu, C., & Lăpădat, C. V. M. (2024). Does Common agricultural policy influence regional disparities and environmental sustainability in European Union Countries? *Agriculture; Basel, 14*(12).

Mazumdar, K. (2003). Do standards of living converge? A cross-country study. *Social Indicators Research, 64*, 29–50.

Meadows, D. H., Meadows, D. L., Randers, J., & Behrens, W. W. (1972). *The limits to growth*. Universal Books.

Meyfroidt, P., & Lambin, E. F. (2011). Global forest transition: Prospects for an end to deforestation. *Annual Review of Environment and Resources, 36*(1), 343–371.

Mignamissi, D., Ekoula, H. W. M. A., & Thioune, T. (2024). Institutions' quality and environmental pollution in Africa. *Environment and Development Economics, 29*(3), 206–233.

Mishra, N., Grima, S., & Ozen, E. (2024). Unpacking the black box: Investigating the role of social protection programmes in promoting decent work and economic growth in low-income countries. *Sustainable Development, 32*(5), 5825–5835.

Modibbo, H., & Inuwa, N. (2020). Health outcomes and economic growth Nexus: Evidence from Nigeria. *Journal of Business and Economic Options, 3*(2), 46–55.

Mosca, I., Van Der Wees, P. J., Mot, E. S., Wammes, J. J., & Jeurissen, P. P. (2016). Sustainability of long-term care: Puzzling tasks ahead for policy-makers. *International Journal of Health Policy and Management, 6*(4), 195.

Moustairas, I., Vardopoulos, I., Kavouras, S., Salvati, L., & Zorpas, A. A. (2022). Exploring factors that affect public acceptance of establishing an urban environmental education and recycling center. *Sustainable Chemistry and Pharmacy, 25*, Article 100605.

Mujtaba, A. (2024). Economic development and social welfare: Assessing the relationship. *Physical Education, Health and Social Sciences, 2*(4), 1–10.

Mukherjee, P. K., Das, B., Bhardwaj, P. K., Tampha, S., Singh, H. K., Chanu, L. D., & Devi, S. I. (2023). Socioeconomic sustainability with circular economy—An alternative approach. *Science of the Total Environment, 904*, Article 166630.

Nguy˜ên, H. V., & Phan, T. T. (2025). Impact of economic growth, international trade, and FDI on sustainable development in developing countries. *Environment, Development and Sustainability, 27*, 4017–4043.

Nguyen, P. D. (2024). Understanding regional development strategies of the European Union and China: Lessons learned for Vietnam. *Sustainable Social Development, 2*(4), 2528.

Nickayin, S. S., Coluzzi, R., Marucci, A., Bianchini, L., Salvati, L., Cudlin, P., & Imbrenda, V. (2022). Desertification risk fuels spatial polarization in 'affected' and 'unaffected' landscapes in Italy. *Scientific Reports, 12*(1), 747.

Ogundipe, A., Alege, P. O., & Ogundipe, O. (2014). Income heterogeneity and environmental Kuznets curve in Africa. *Journal of Sustainable Development, 7*(4), 165–180.

Oketch, M. O. (2006). Determinants of human capital formation and economic growth of African countries. *Economics of Education Review, 25*(5), 554–564.

Osberg, L., & Sharpe, A. (2005). How should we measure the "economic" aspects of well-being? *Review of Income and Wealth, 51*(2), 311–336.

Ossewaarde, M., & Ossewaarde-Lowtoo, R. (2020). The EU's green deal: A third alternative to green growth and degrowth? *Sustainability, 12*(23), 9825.

Overmars, K. P., Stehfest, E., Tabeau, A., van Meijl, H., Beltrán, A. M., & Kram, T. (2014). Estimating the opportunity costs of reducing carbon dioxide emissions via avoided deforestation, using integrated assessment modelling. *Land Use Policy, 41*, 45–60.

Papamichael, I., Voukkali, I., Loizia, P., Stylianou, M., Economou, F., Vardopoulos, I., Klontza, E. E., Lekkas, D. F., & Zorpas, A. A. (2023). Measuring Circularity: Tools for monitoring a smooth transition to Circular Economy. *Sustainable Chemistry and Pharmacy, 36*, Article 101330.

Parrique, T., Barth, J., Briens, F., Kerschner, C., Kraus-Polk, A., Kuokkanen, A., & Spangenberg, J. H. (2019). Decoupling debunked. In *Evidence and arguments against green growth as a sole strategy for sustainability.* A study edited by the European Environment Bureau EEB, Bruxelles.

Permanyer, I., & Scholl, N. (2019). Global trends in lifespan inequality: 1950–2015. *PLoS ONE, 14*(5), Article e0215742.

Petrakos, G., Rontos, K., Salvati, L., Vavoura, C., & Vavouras, I. (2024). Income inequality in the over-indebted Eurozone countries and the role of the excessive deficit procedure. *Open Economies Review, 35*(2), 301–322.

Quaranta, G., Salvia, R., Salvati, L., Paola, V. D., Coluzzi, R., Imbrenda, V., & Simoniello, T. (2020). Long-term impacts of grazing management on land degradation in a rural community of Southern Italy: Depopulation matters. *Land Degradation and Development, 31*(16), 2379–2394.

Raghupathi, V., & Raghupathi, W. (2020). The influence of education on health: An empirical assessment of OECD countries for the period 1995–2015. *Archives of Public Health, 78*, 1–18.

Rahman, M. M., & Alam, K. (2021). Exploring the driving factors of economic growth in the world's largest economies. *Heliyon, 7*(5), Article e07109.

Ramani, K. V., & Mavalankar, D. (2006). Health system in India: Opportunities and challenges for improvements. *Journal of Health Organization and Management, 20*(6), 560–572.

Ranis, G., Stewart, F., & Ramirez, A. (2000). Economic growth and human development. *World Development, 28*(2), 197–219.

Raworth, K. (2017). *Doughnut economics: Seven ways to think like a 21st-century economist*. Chelsea Green Publishing.

Redclift, M. (2005). Sustainable development (1987–2005): An oxymoron comes of age. *Sustainable Development, 13*(4), 212–227.

Ribeiro Hoffmann, A., Sandrin, P., & Doukas, Y. E. (2024). Climate change in regional perspective. European Union and Latin American initiatives, challenges, and solutions. Cham, Switzerland: United Nations University Series on Regionalism, Vol. 27, Springer Nature.

Riouch, A., Benamar, S., Ezzeri, H., & Cherqi, N. (2024). Assessing Student perceptions of pollution and management measures related to COVID-19 vaccination tools in Morocco. *Pakistan Journal of Life and Social Science, 22*(2), 1–22.

Rodrigo-Comino, J., Salvia, R., Quaranta, G., Cudlín, P., Salvati, L., & Gimenez-Morera, A. (2021). Climate aridity and the geographical shift of olive trees in a Mediterranean Northern Region. *Climate, 9*(4), 64.

Russo, G., Silva, T. J., Gassasse, Z., Filippon, J., Rotulo, A., & Kondilis, E. (2021). The impact of economic recessions on health workers: A systematic review and best-fit framework synthesis of the evidence from the last 50 years. *Health Policy and Planning, 36*(4), 542–551.

Sadik-Zada, E. R., & Gatto, A. (2023). Grow first, clean up later? Dropping old paradigms and opening up new horizons of sustainable development. *Sustainability, 15*(4), 3595.

Sallustio, L., Pettenella, D., Merlini, P., Romano, R., Salvati, L., Marchetti, M., & Corona, P. (2018). Assessing the economic marginality of agricultural lands in Italy to support land use planning. *Land Use Policy, 76*, 526–534.

Salvati, L. (2016). The dark side of the crisis: Disparities in per capita income (2000–12) and the urban-rural gradient in Greece. *Tijdschrift Voor Economische En Sociale Geografie, 107*(5), 628–641.

Salvati, L. (2025). *Environmental sustainability and global change*. Elsevier.

Salvati, L., & Carlucci, M. (2014). A composite index of sustainable development at the local scale: Italy as a case study. *Ecological Indicators, 43*, 162–171.

Salvati, L., Ciaschini, C., Postigliola, M., Ermini, B., & Vardopoulos, I. (2025). Contextual differences, contrasting trends? Sustainable development, economic performances, and divergence in land resource depletion. *Environmental Sustainability and Global Change* (pp. 107–114). Elsevier.

Salvati, L., & Zitti, M. (2009). Substitutability and weighting of ecological and economic indicators: Exploring the importance of various components of a synthetic index. *Ecological Economics, 68*(4), 1093–1099.

Sateriano, A., D'Agata, A., & Vardopoulos, I. (2024). *Socio-environmental change: The path toward sustainability*. Nova Science Publishers.

Sharma, H. B., Vanapalli, K. R., Samal, B., Cheela, V. S., Dubey, B. K., & Bhattacharya, J. (2021). Circular economy approach in solid waste management system to achieve UN-SDGs: Solutions for post-COVID recovery. *Science of the Total Environment, 800*, Article 149605.

Siddiqi, S., Aftab, W., Raman, A. V., Soucat, A., & Alwan, A. (2023). The role of the private sector in delivering essential packages of health services: Lessons from country experiences. *BMJ Global Health, 8*(1), Article e010742.

Soubbotina, T. P. (2004). *Beyond economic growth: An introduction to sustainable development*. World Bank Publications.

Spaiser, V., Ranganathan, S., Swain, R. B., & Sumpter, D. J. T. (2016). The sustainable development oxymoron: Quantifying and modelling the incompatibility of sustainable development goals. *International Journal of Sustainable Development & World Ecology, 24*(6), 457–470.

Spence, M. (2024). *The next convergence: The future of economic growth in a multispeed world*. Macmillan+ ORM.

Spilioti, N., & Anastasiou, A. (2024). European structural and investment funds (ESIFs) and regional development across the European Union (EU). *Journal of Risk and Financial Management, 17*(6), 228.

Stojkoski, V., Koch, P., & Hidalgo, C. A. (2023). Multidimensional economic complexity and inclusive green growth. *Communications Earth and Environment, 4*(1), 130.

Sun, X., Qing, J., Shah, S. A. A., & Solangi, Y. A. (2023). Exploring the complex nexus between sustainable development and green tourism through advanced GMM analysis. *Sustainability, 15*(14), 10782.

Tabutin, D., Masquelier, B., Grieve, M., & Reeve, P. (2017). Mortality inequalities and trends in low-and middle-income countries, 1990–2015. *Population, 72*(2), 221–296.

Tamazian, A., & Rao, B. B. (2010). Do economic, financial and institutional developments matter for environmental degradation? Evidence from transitional economies. *Energy Economics, 32*(1), 137–145.

Tödtling, F., & Trippl, M. (2005). One size fits all? Towards a differentiated regional innovation policy approach. *Research Policy, 34*(8), 1203–1219.

Tsangaris, S., Xepapadeas, A., Yannacopoulos, A. N., & Salvati, L. (2024). Spatial externalities, R&D spillovers, and endogenous technological change. *Regional Science and Urban Economics, 109*, Article 104055.

Vardopoulos, I., & Konstantinou, Z. (2017). Study of the possible links between CO_2 emissions and employment status. *Sustainable Development, Culture, Traditions Journal, 1*(B), 100–112.

Vardopoulos, I. (2018). Multi-criteria decision-making approach for the sustainable autonomous energy generation through renewable sources. Studying Zakynthos Island in Greece. *Environmental Management and Sustainable Development, 7*(1), 52–84.

Vardopoulos, I. (2023). Adaptive reuse for sustainable development and land use: A multivariate linear regression analysis estimating key determinants of public perceptions. *Heritage, 6*(2), 809–828.

Vardopoulos, I. (2025). Cities in flux: Waste, the shadow of circularity, and the (elusive) promise of sustainability. *Sustainability, 17*(2), 705.

Vardopoulos, I., Abeliotis, K., & Lasaridi, K. (2025). A systematic informetric analysis and literature review of food waste quantification studies in the food service industry. *Sustainability, 17*(1), 103.

Vardopoulos, I., Giannopoulos, K., Kalli, S., & Salvati, L. (2024). From tech-centric to human-centric smart tourist destinations: Innovations in real estate, urban planning, and sustainability. *Socio-enviromental change: The path toward sustainability* (pp. 65–96). Nova Science Publishers.

Vardopoulos, I., Vannas, I., Xydis, G., & Vassiliades, C. (2023). Homeowners' perceptions of renewable energy and market value of sustainable buildings. *Energies, 16*(10), 4178.

Vermeulen, B., Kesselhut, J., Pyka, A., & Saviotti, P. P. (2018). The impact of automation on employment: Just the usual structural change? *Sustainability, 10*(5), 1661.

Vinci, S., Vardopoulos, I., & Salvati, L. (2023). A tale of a shrinking City? Exploring the complex interplay of socio-demographic dynamics in the recent development of Attica Greece. *Cities, 132*, Article 104089.

Wang, J., Wang, B., Dong, K., & Dong, X. (2022). How does the digital economy improve high-quality energy development? The case of China. *Technological Forecasting and Social Change, 184*, Article 121960.

Wang, J., Yang, J., & Yang, L. (2023). Do natural resources play a role in economic development? Role of institutional quality, trade openness, and FDI. *Resources Policy, 81*, Article 103294.

Wang, Y., He, Y., & Gao, X. (2025). Synergizing renewable energy and circular economy strategies: Pioneering pathways to environmental sustainability. *Sustainability, 17*(5), 1801.

Ward, J. D., Sutton, P. C., Werner, A. D., Costanza, R., Mohr, S. H., & Simmons, C. T. (2016). Is decoupling GDP growth from environmental impact possible? *PLoS ONE, 11*(10), Article e0164733.

Wenjuan, S., & Zhao, K. (2023). Balancing fiscal expenditure competition and long-term innovation investment: Exploring trade-offs and policy implications for local governments. *PLoS ONE, 18*(11), Article e0293158.

Yusuf, N., & Nasrulddin, V. (2024). Significance of international trade and national GDP as two integral components of sustainable economic development in Saudi Arabia. *Journal of the Knowledge Economy, 15*(1), 2298–2317.

Zambon, I., & Salvati, L. (2019). Metropolitan growth, urban cycles and housing in a Mediterranean country, 1910s–2010s. *Cities, 95*, Article 102412.

Zhan, J. X., & Santos-Paulino, A. U. (2021). Investing in the sustainable development goals: Mobilization, channeling, and impact. *Journal of International Business Policy, 4*(1), 166.

Zhang, C., Waris, U., Qian, L., Irfan, M., & Rehman, M. A. (2024). Unleashing the dynamic linkages among natural resources, economic complexity, and sustainable economic growth: Evidence from G-20 countries. *Sustainable Development, 32*(4), 3736–3752.

3 Age Dependency Ratio

Introduction

In 1950, the world population structure consisted of an increasing number of births and a higher risk of death throughout life (Zhang et al., 2015). Consequently, the level of new borns was higher compared with the number of old people. In the last years, population structures significantly changed (Loichinger & Skirbekk, 2016), with a correspondent decrease in children mortality (Abbas et al., 2024). Different considerations should be made comparing outcomes of 2018 to the future projections of 2100. According to these projections, indeed, the number of new born will sharply decrease at the end of the century (Hammer et al., 2015; Maestas & Zissimopoulos, 2010; Preston, 1984). As global health is improving and mortality is decreasing, life expectancy will be higher, and death rate will be even lower (Lee, 2002), while lower income countries converge to a lower median age (Uhlenberg, 2009).

Since an efficient economic structure of societies depends on the age structure of its population (Oppenheimer, 1994), the age dependency ratio acquires crucial relevance in determining the distribution between the working-age population and the 'dependent' population (Slaveva, 2023). The indicator is crucial to achieve the 2030 Sustainable Development Goals of the United Nations (Leal Filho et al., 2023; Rai et al., 2019; Zammitti et al., 2024), particularly, SDG8 connected to the work environment, but also, the two SDGs connected to education (SDG4) and gender equality (SDG5).

Literature Review

The Age Dependency Ratio (hereafter 'ADR') is generally used to summarize age distributions, specifically indicating whether a population is 'too young' or 'too old'. The dependent population is defined as the share of the population economically inactive and not earning labour income (Ram, 1982). It includes children and the elderly, typically relying on others for their consumption and goods requirements (Leff, 1969). Within this context, a higher ADR means a greater burden on working people in specific geographic areas (Bernheim, 1987). Consequently, a wide percentage of economically 'dependents' relative to the working-age population negatively impacts labour productivity, capital formation and saving rates (Choudhry et al., 2016). A specific area's dependency ratio derives from the aggregated quantification of the two 'dependent' populations then divided by the total number of producers (Darby & Melitz, 2008; Grant & Benton, 2000; Paterson et al., 2004). These results direct affect financial expenditures on items like social security (Bellettini & Ceroni, 2000; Ehrlich & Zhong, 1998).

C. Ciaschini and L. Salvati, *Sustainable Development*, Sustainable Development Goals Series,
https://doi.org/10.1007/978-3-032-12559-0_3

Economic Policy and Welfare Issues

The economic aspects connected to ADR trends are crucial to understand challenges and development opportunities of a given country (An & Jeon, 2006; Disney, 1996; Kulik et al., 2014). Environments characterized by a high ADR usually place a weight on the welfare system and on public resources dedicated to social protection, as retirement plans, healthcare assistance, and social services (Galasso & Profeta, 2004). Public resources should be allocated more effectively through policies containing expenditure or reform pensions, and the job market (French & Jones, 2012; Salvati & Zitti, 2009; Vardopoulos et al., 2021). Within this framework, the sustainability of public finance would then require long-term strategies to enable the provision of essential services without compromising fiscal stability (Andrews & Cingano, 2014; Egidi et al., 2020; Salvati & Zitti, 2008). A high ADR also affects economic growth. An increasing share of the inactive population, in turn, implies a reduction in the number of people contributing to production and innovation, negatively affecting Gross Domestic Product (GDP) and economic dynamics at large (Shabat, 2020). Workforce decrease could also imply a lack of competences and productive capacities, to be addressed through policies of professional formation (Kasnauskienė & Andriuškaitė, 2017) and incentives for qualified immigration (Marois et al., 2022).

To this purpose, digitalisation and automation are revealed as complementary instruments to optimise labour efficiency and mitigate the effects of demographic ageing (Kim et al., 2013). Specific strategies should be designed to encourage the working participation of traditionally more vulnerable groups, i.e. women and the elderly (Inman, 1982). Support policies to conciliate work and family life, as assistance to childhood and flexibility in work hours, can contribute to increasing the participation rate of the active population (Gonzalez-Eiras & Niepelt, 2012; Holzmann, 2013; Pezzulo et al., 2017). In addition, fiscal incentives for firms hiring specific working groups can preserve a balance between the demand and supply of work (Muszyńska & Rau, 2012). Globally, pension reforms, to ensure long-term sustainability, should be supported by measures enhancing productivity and technological innovation, for a more robust economic growth, even with a progressively ageing population (Wang & Wang, 2025).

In conclusion, relevant investments in policies favouring education and formation of human capital are crucial to improve productivity and effectively respond to demographic changes (Wei & Hao, 2010). A more inclusive and resilient economy requires an integrated approach, considering both demographic dynamics and sustainable development strategies (Disney, 2004; Hock & Weil, 2012; Phurisikul & Phuakkong, 2025). Moreover, welfare represents a crucial element to support the needs of ageing populations. An effective welfare system ensures that people, regardless of their age, can access essential services as healthcare assistance, social providence, housing assistance, and social services. The sustainability of welfare also depends on the balance between available resources and increasing requirements, to be attained through careful planning and reforms of public policies (Salvati & Carlucci, 2011).

Within this context, healthcare assistance is a crucial pillar of life quality, especially for the elderly (Daly & Lewis, 2000). The network of public and private healthcare services should be able to respond to the needs of an older population, providing preventive care, rehabilitative care, and long-term care (Pfau-Effinger, 2012). The promotion of healthy lifestyles, prevention of diseases, and universal access to care are key elements to reduce the impact of chronic diseases and disability on society (Madhavan et al., 2009). Moreover, public welfare should also design programmes of social assistance and support to relieve many people and families from economic difficulties, thus favouring social inclusion (Hock & Weil, 2012). Supports as pensions, family allowance, and services of domiciliary assistance should support the creation of a more inclusive social path (Gough, 1996; Mbata et al., 2024; Preston, 2007).

A further relevant aspect relates to the policies of integration and support of work, inserted within a framework of welfare aimed at favouring the independence and the inclusion of all population classes (Gerland et al., 2014). Programmes of professional formation, incentives to encourage assumptions, and policies of conciliations between work and family life are useful tools to reduce unemployment and support stable and quality employment (D'Agostino & Regoli, 2013; Ludwig et al., 2012; Pecchenino & Utendorf, 1999). In addition, welfare also deals with the safeguard of workers in situations of evident vulnerability, ensuring rights and services to face the issues connected to diseases, disabilities, or changes in job conditions (Hadley et al., 2011). This care for the welfare of workers favours not only economic growth, but also a more equitable and stable social climate (Thinley, 2021). In general, a sustainable welfare system requires careful management of resources and a continuous interaction between institutions, social partners, and citizens (Gough et al., 1997; Harper, 2014; Santacreu, 2016). Policies favouring innovation in social services, efficiency of resources, and inclusion of new technologies, should improve the provision of services (Minas, 2016). In an era of economic crisis and accelerated demographic transitions (Razin et al., 2002) the adaptability of the welfare model would ensure a more equitable and sustainable future for all generations (Esping-Andersen, 2000).

Healthcare Issues

ADR provides a crucial measure to understand how the composition of the population affects the capacity of the healthcare system in responding to the needs of the population itself (Cruz & Ahmed, 2018; Zhao, 2014). A high ADR indicates a significant proportion of active people under the economic viewpoint, as children and the elderly, with respect to the working-age population (Inácio et al., 2018). This situation implies different effects on planning and management of healthcare resources (Ilan, 2021). For instance, in areas with high ADR, an increase in demand for long-term assistance services, palliative care, and social support services for the elderly was observed (Jokinen et al., 2013). These services require specialized resources and a capillary organization, capable of putting together public and private healthcare systems (Singh & Prakash, 2010).

The necessity of ensuring quality assistance and a population in rapid ageing implies an increase in healthcare costs, which should be supported through private or public resources (Sekhri et al., 2011). An increase in ADR can also affect the sustainability of social providence systems and healthcare assistance (Iwański, 2024). With a wider share of retired people and inactive people, fiscal entries dedicated to finance healthcare and social services tend to decrease, while expenditures increase (Dao, 2012; Senbekov et al., 2020; Wolff et al., 2020). This imbalance lead to an increasing pressure on public financing systems, making interventions of reform and resources optimisation more and more urgent (Ezeh et al., 2012). Healthcare policy planning should, therefore, carefully consider the demographic trends, forecasting prevention strategies, and promotion of healthcare that is able to contain the costs and improve the life quality of more vulnerable demographic classes (Aday, 1994; Dang et al., 2020; Prince et al., 2015).

In addition, population ageing, reflected in a higher ADR, has implications for research and development in the healthcare sector (Lee & Yoon, 2021). Investments in technological innovations, as digital therapies and artificial intelligence, should increase the effectiveness of assistance and reduce expenditures (Laurenza et al., 2018). Telemedicine, for instance, allows for remote monitoring of old patients, improving the quality of care, and reducing the frequency of hospital visits (Da Silva et al., 2021; Feijóo et al., 2020; Mechanic & Tanner, 2007). In parallel, programmes of prevention for older people should maintain long-lived people in good health and reduce dependence on intensive healthcare services (Williams et al., 2016). In summary, ADR is an effective tool to plan effective and sustainable public health policies over time (Suzman et al., 2015). An effective connection between population ageing and the need for efficient, equitable, and universally accessible healthcare systems (e.g. Cristea et al., 2020), should help face the challenges in a world of rapid demographic and epidemiological change (Khan et al., 2024).

Empirical Findings

The present study investigates trends over time in the Adolescent Fertility Rate (ADR) based on the World Bank database of World Development Indicators (WDI). The index is analysed for 184 countries over a time horizon of 60 years, from 1960 to 2020, and aims at understanding the negative impact of population age structure on communities and its path of development over time. Considering the spatial distribution of ADR across world countries in 1960, we can note how the age dependency ratio of most countries in the Southern hemisphere was close to 100%, since the number of 'dependents' correspond to the number of working-age people. In four countries, namely Paraguay, Suriname, Kenya and the Philippines, the age dependency ratio lays above 100%, meaning more dependents than working-age people. The spatial distribution of the same indicator for 2020 highlights a marked change, since the dependent population stabilized to 50–60% of the working-age group in most countries. Nevertheless, a higher ratio was observed in sub-Saharan African countries such as Niger and Mali, with some exceptions in countries as Qatar, maintaining a low ADR since 1990, and dropping to 18% in 2020.

A proper analysis of latent trends in ADR on a given area, however, should take two main concepts in a due consideration. On the one side, a distinction should be made between the ADRs of young and elderly groups. On the other side, the relationship between labour force participation rates and age groups should be also considered. The first distinction underlines the relevance of ADR in long-term regional development. Different age groups differently impact on both the environment and infrastructure. By decomposing the (total) dependency ratio into (i) youth dependency ratio and (ii) old-age dependency ratio, a particularly high value of the young-age dependency ratio, especially in sub-Saharan Africa, can be noted. Some countries in this area have almost the same number of young people over the working-age population, resulting in very young populations.

Specifically, in 2020, Mali had an age dependency ratio of 98% of which only 5% belongs to the elderly population. Similarly, Nigeria performed an ADR of 85%, with 80% for young and 5% for elderly groups. The youth dependency ratio is lower throughout higher income countries characterized by lower fertility rates. In this perspective, Singapore has the lowest youth dependency rate, equal to 17% in 2020. This country has a high GDP per capita, a low total fertility rate and a long-life expectancy, at least in 2020. In general, this framework of low dependency ratio would represent a relevant issue in the future, particularly in case of low fertility. In the future scenarios, indeed, 'producing' populations will retire and should be replaced by younger workforce.

Countries with the highest dependence on old age are in Europe, North America and East Asia. According to this index, Italy is the opposite of sub-Saharan African countries (such as Niger and Mali). Since the late 1970s, Italy had an age dependency ratio of 57%, more than half of the recorded rate being the specific contribution of elders. As already mentioned, a high dependency ratio raises major issues since a wide amount of public expenditure is employed for health, social security, and education, more required by the dependent population. Fewer working-age people mean fewer people in charge of schools, old-age pensions, disability pensions and other forms of assistance for dependent population, often the most vulnerable members of society (Prskawetz & Sambt, 2014).

The second aspect relates to age and job participation. Conventionally, a direct relationship is assumed between age and work productivity. As a consequence, the worker belongs to the age group 15–64, while the dependent rests in the other age groups. Nevertheless, ADR ignores that many over 65 s are not necessarily dependent (for instance, an increasing percentage of them work), and that many of those included in 'working age' are not actually employed. Cultural issues such as an acceptable retirement age, delaying work for education and the role of women in the workforce vary greatly among countries and over time (Salvati & Carlucci, 2014). All these factors hinder the adoption of ADR when comparing economic dynamics, e.g. the volatile balance between workers and non-workers (Skirbekket et al., 2022). Therefore, measures of dependency only based on chronological age could be misleading.

In general, a descriptive analysis show that high-income countries have high old-age dependency ratio compared to the other parts of the world (Tatem et al., 2014). The expected increase in the elderly dependency ratio in these countries will strongly affect their economic systems (Tsai et al., 2000). These young dependents will, then, enter the working-age bracket, in turn implying a significant increase in the share of working-age people in the coming decades. As a result, the ADR will decrease over the years. A working population in charge of fewer dependent people leads to a relevant acceleration in economic growth, thus determining a demographic dividend. In literature, studies on the demographic dividend are largely based on comparative analyses of age dependency ratios. Bloom et al. (2011) argued how the 'demographic dividend' is the potential for economic growth, deriving from the change in the age structure of a given population, particularly in cases of a working-age population (15 to 64 years old) greater than the 'dependents' (up to 14 years and over 65).

This potential can be enormous, as long as supportive policies are in place and invest in human capital, particularly in young people, with substantial and strategic measures. To record a demographic dividend, the country should first enter a process of demographic transition (Salvati & Carlucci, 2011). In the early stages of the demographic transition, infant mortality rates decrease, thanks to interventions promoting drinking water and sanitation. As a result, healthier children are more likely to survive. The immediate effect of the decrease in infant mortality is an increase in the number of children. In these cases, families employ money for their children, taking away resources likely to be invested in small businesses, expanding agricultural production, or for other purposes (Fehr et al., 2008). Governments should, then, invest more resources for healthcare and primary education, often hindering economic development. Increased chances of children's survival induce parents to opt for a smaller family, thus contributing to a gradual reduction in fertility rates, the next stage of demographic transition (Zhang & Zhang, 2005).

Fewer children mean more investments improving the income of the whole family in the long run. Those born in the first stage of transition become adults and start working (Ashraf et al., 2013). When this growing percentage of the working population starts to perceive a salary, the country's per capita income typically increases. In addition, the decrease in fertility rates encourages women to join the labour force. Bloom et al. (2011) described the decline in fertility as a powerful driver towards changes in the age structure of a given population, with important economic repercussions. According to Lee and Mason (2006), the period in which such a demographic dividend can be recorded is very long, even five decades or more (Mehta & Awasthi, 2025). Nevertheless, in latest years, the decrease in fertility will lead to a reduction in the growth rate of the workforce (Navaneetham & Dharmalingam, 2012), while the continued advances in medicine will lead to an increase in life expectancy and a corresponding growth in the elderly population, e.g. as in the case of Japan (Miles, 2023).

The demographic dividend is an opportunity for rapid economic growth (Chandrasekhar et al., 2006), but its achievement requires adequate policies and investments in the economic and social fields (Taneja, 2018). The size of this dividend therefore depends on the political and economic response of governments (Brown et al., 2008). New generations entering the labour market represent a great economic potential (Abdullah & Mansoor, 2024), only if families and governments have invested adequately in their health and education, thus creating new opportunities (Frankowska et al., 2015; James, 2011; Varsha Pramod & Ramachandran, 2023).

Descriptive Statistics

The empirical results reported in Table 3.1 reflect the specific trend in the age dependency ratio for each decade of the analysed period, i.e. between 1960 and 2020. The mean values remark a progressive decrease in the studied ratio; nevertheless, some peculiarities emerge, as for 1960 and the peak of 1970. Exceptions are highlighted not only for the mean values, but also for other statistical metrics summarized in the table. This aspect emerges because the age dependency ratio considers both the youth age dependency ratio and the old age dependency ratio, performing different values for affluent and emerging countries. In literature, a high age dependency ratio is connected with two main characteristics. On the one side, it reports a high level of young population—a typical feature of developing countries—in turn characterized by high fertility rates. On the other hand, it could be connected with high levels of older population, usually observed in affluent countries, with a better quality of life and higher levels of income.

In general, literature highlights different paths in the countries' development. With reference to a given year, some countries have registered an increase in fertility rates and, in turn, in births' number, while others reported higher levels of the elderly population. As a result, each country performs heterogeneous values of the age dependency ratio, accounting for both the elderly and young population. The minimum observed value within the statistical distribution of ADR reduced from 1960 to 1970, with a recovery between 1970 and 1980, a significant decrease from 1980 to 2010, and a new increase in 2020. The decrease in ADR raises expectations of future increases in working-age population over dependents. Hence, a demographic dividend is expected in countries previously characterized by youth dependency, now constituting the workforce. Maximum values reflect the tendency observed in affluent countries, characterized by a relevant share of young population. In most cases, disadvantaged economies may perform the highest values of ADR compared with developed countries, only recently recording a significant increase in elderly. Moreover, in affluent countries, old age dependency ratio still remains the lowest share of the total ADR.

Conversely, emerging economies perform high values of youth dependency ratio connected with the maximum values of total ADR. Average values seem to be more precise, with lower standard deviations in the first decades. In subsequent years, increasing standard errors highlight a greater dispersion of the observed values around the mean. The median value of the index decreases throughout decades, with the sole exception of 1960 and 1970, in correspondence with the so-called 'demographic boom', characterized by an increase in the youth dependency ratio compared to total ADR. The median value increases also in 2020, for a slightly risen old-age dependency ratio of high-income countries with older populations. The specific trend over time in the 25th percentile denotes that, over the analysed time horizon, the first quarter of the investigated countries show a

Table 3.1 Descriptive statistics representing the distribution of age dependency ratio across world countries by decade

Metric	1960	1970	1980	1990	2000	2010	2020
Arith. mean	79.4	82.2	77.4	73.1	67.2	60.1	58.9
Minimum	47.1	44.9	46.3	37.1	33.5	15.7	18.1
Maximum	117.5	120.5	112.6	117.9	110.8	110.7	109.5
Norm. range	0.9	0.9	0.9	1.1	1.2	1.6	1.6
Median	81.6	85.6	81.1	73.4	62.5	53.0	55.2
25th percentile	70.1	67.1	58.6	52.2	49.6	46.7	48.7
75th percentile	90.0	95.3	92.5	91.5	82.9	72.0	67.9
Skewness	− 0.3	− 0.3	− 0.1	0.1	0.4	0.8	0.8
Kurtosis	− 0.5	− 0.8	− 1.2	− 1.4	− 1.0	− 0.3	0.5

Source Authors' elaboration of World Bank indicators

value of the indicator between 46 and 70. From the 75th percentile onwards, values lay constantly above the average. This means that the highest percentage of the dependents are concentrated in a small number of states. Looking at the skewness, the observed distribution is mostly symmetrical. Negative values for the skewness, recorded in the decades from 1960 to 1980, indicate that data are left skewed, while, for the decades after 1980, positive values indicate a right skewness. The dataset shows a low level of kurtosis and, in turn, light tails or no outliers. In line with the literature, the peaks in the 1960s, 1990s and late 2010s, account for the interaction of ADR with several different indicators, as fertility rates, life expectancy and per-capita Gross Domestic Product (GDP).

Correlation Analysis

Figure 3.1 illustrates the evolution of Pearson's and Spearman's correlation coefficients from 1961 onwards. The chart shows how both coefficients assume positive values close to 1, moving from year to year. Specifically, the results of Spearman's correlation coefficient almost resembles those obtained with Pearson's correlation index. Rapid population ageing in most affluent countries will continue to raise the age dependency ratio. For instance, in 1961, Lithuania had a total age dependency ratio of 54% (made up of 42% young dependents and only 12% elderly). In 2020, the ADR composition of Lituania, equal to 56%, has drastically changed. More than half of the dependents belongs to the elderly, 33%, and the remaining share, equal to 24%, belongs to the young population. In this country, the interaction between increased life expectancy, declining birth rates, and massive emigration flows, is a crucial feature in defining ADR. Another example is Norway where a total ADR of 58% was recorded for 1961, further deacreasing only by 5 percentage points in 59 years, and reaching, in 2020, the value of 53%. A huge increase in the dependency ratio could negatively affect economic growth.

Examples of reverse patterns include Arab countries, as Morocco. These countries were characterized by high fertility in the 1960s, and marked population growth rates. No fundamental decrease in fertility was seen until 1965. These new trends have been also encouraged by the family planning programmes launched in the 1970s (Youssef and Elden, 2018). Furthermore, economic crisis and the relevant military expenditure connected to the war in Sahara widely affected fertility rate and the economic growth as a whole. These factors largely affected ADR, drastically reduced compared to the early 1960s. Data seem to confirm this dynamics; in 1961 the dependency ratio between the total age in Morocco was 94%, among which a large part was due to young population

Fig. 3.1 Trends over time in parametric (Pearson—black line) and non-parametric (Spearman—grey line) correlation coefficients within the (−1) lagged time series of the age dependency ratio across world countries. *Source* Authors' elaboration of World Bank indicators

(equal to 88%). This value has been drastically reduced to 52% in 2020, approximately half of the value recorded in 1961.

Multidimensional Analysis

Principal component analysis (PCA) extracted two main axes with the aim at summarizing the entire data matrix. Component 1 (PC1) resulted as the dominant dimension, accounting for 75% of the total variance, while Component 2 (PC2) explained only 19%, bringing the cumulative variance explained by the first two components to 94%. The remaining components have explained a minimal variance. PC1 captures the most significant demographic trends affecting economic growth, including the impact of age dependency ratios. Countries with low dependency ratios (e.g. those benefiting from a demographic dividend with a large working-age population) likely align closely with PC1. These countries primarily belong to East Asia during rapid growth periods, or include some emerging economies in Africa and South Asia. Countries with favourable dependency ratios (basically, with low dependency) benefit from increased savings, investments, and productivity. These factors are strongly correlated with PC1. Regions with high dependency ratios (youth or elderly) face lower economic growth due to higher burdens on the working-age population. These effects may appear in PC2, underscoring the strong relationship between demographic structures and economic growth. As a summary, PC1 identifies and describe spatio-temporal trends in dependency ratios and their impact on economic performances, while PC2 reflects secondary differences between youth and elderly dependency across regions.

Figure 3.2 allows identifying countries with different characteristics connected with age dependency, providing a useful tool for economic, demographic and political interpretations. Points right from the origin represent countries with higher values of variables strongly affecting Component 1, namely countries with high rates of age dependency characterized by a young population, or by a strong dependency from elderly. Points left from the origin represent countries with lower values for dominant variables in Component 1. Positive values of Component 2 define countries featuring some (even irrelevant) peculiarities, as a high level of economic development, a higher life expectancy, or social policies affecting the dependency ratio. Low points relate to countries with less developed policies or differ-

Fig. 3.2 Results of a principal component analysis (PCA) as illustrated in a biplot reflecting the distribution of variables' (i.e. years) loadings and rows (i.e. countries) scores along the first two axes. *Source* Authors' elaboration of World Bank indicators

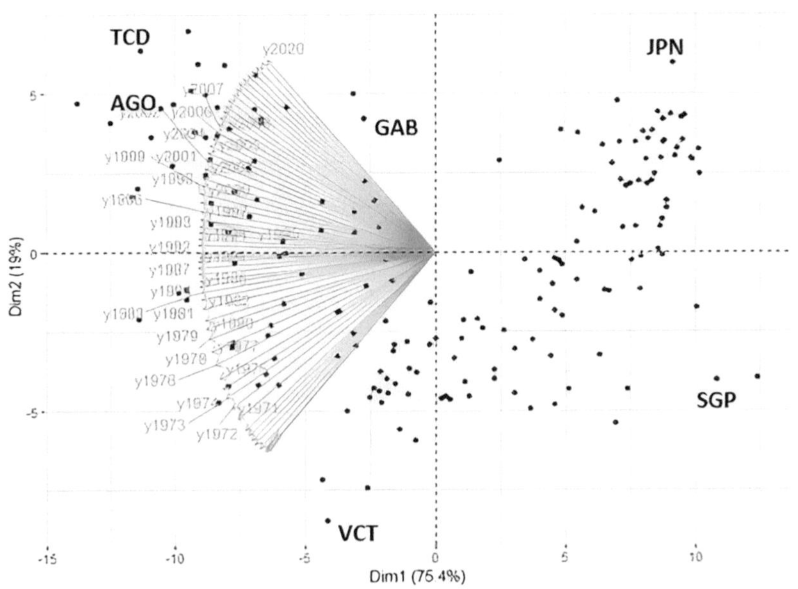

ences in the demographic structure affecting the dependency indicator. Countries located near to the origin, as Gabon (GAB), present a balanced population in terms of youth (active population), and elderly, or with an average level of economic and demographic development.

Figure 3.2 also highlights isolated clusters, i.e. St. Vincent and the Grenadines (VCT) in the bottom left of the plot, Chad (TCD) at the top left, Japan (JPN) in the top right, Singapore (SGP) in the bottom right, and Angola (AGO) at the top. St. Vincent and the Grenadines (VCT) perform negative values for both Components 1 and 2, being a country with a relatively young population and lower economic development, where the age dependency ratio is less significant compared with other variables. Chad (TCD) is a country with a mature yet active population, featuring low birth rates. Japan (JPN) represents a country with a wide share of elderly population (i.e. reflecting high dependency among the elderly), and with a developed economy as well as advanced social policies supporting the demographic burden. Singapore (SGP) is a country with high birth rates and a strong dependency among the young, but with possibly limited resources to support social development. Angola (AGO) is oriented towards the arrow tips indicating a strong correlation with dominant demographic and socioeconomic variables. This country is highly affected by age dependency, with either a very young or very old population, marked by exceptional demographic dynamics.

Discussion

This work analyses the age dependency ratio for different countries in the world, throughout a time horizon of 60 years (from 1960 to 2020). The age dependency ratio measures the age structure of the population. It indicates the ratio between the number of individuals who may be 'dependent' on the support of others for their daily life (young and old) and the number of those able to provide such financial support. A low value of dependency ratio means enough working people to support the 'dependent' population. This context allows for better pensions and better healthcare for citizens (Bidisha et al., 2019). A higher ratio indicates a relevant financial burden for workers, and possible political instability. The analysis highlights a close relationship between the indicator with various factors, e.g. population structure, Gross Domestic Product (GDP) per capita, life quality and life expectancy, fertility rates, and the overall level of socioeconomic development in a given area (e.g. Wilson et al., 2017).

The age structure of a population is relevant for understanding long-term socioeconomic dynamics. Over the decades, the age structure of a population has changed relevantly. In the first decades of observation, there was a high number of births and a low life expectancy. As a result, there have been many new born compared to the number of older people. On the other hand, the population structure has recently recorded more people of all ages added to the world population for each decade. Nevertheless, a comparison between the first and last period of the historical series under scrutiny highlights, for the developing countries, a still strong increase in the number of new born, thus involving a high youth dependency ratio (Ahmed, 2014).

Conversely, this pattern was rapidly changing in affluent countries. As global health is improving and mortality is decreasing, people living today are expected to live longer than previous generations. Nonetheless, these improved life conditions coexist with a decrease in the number of new born. The study also highlights the strong connection between the income of the analysed countries and the age dependency index. Within this context, two different paths emerge, i.e. higher-income countries (such as North America, Europe and East Asia) and lower-income countries (Barrientos et al., 2003). The former tend to have a higher average age and, thus, a high old age dependency ratio; the latter tend to have a lower average age with a high youth age dependency ratio.

In this sense, age dependency ratio defines which countries have a 'too young' or 'too old' population. Moreover, the total age dependency ratio allows the understanding of the decomposition of this indicator between young and old population (Beer & Boswell, 2002). Needs, behaviours, and future pathways for young and old populations are very different (Admassie, 2002). This may affect national planning, from education and healthcare services to labour supply, savings rates and pensions. A more accurate analysis cannot demise the empirical results of a comparative study of the two 'dependent' age groups individually (elderly population and young population).

This is fundamental, because otherwise it would create confusion and an inadequate approximation of the situation experienced by the country in that historical period (Jeetoo & Jaunky, 2023). The analysis allows for the identification of affluent and disadvantaged economies and, in turn, their total age dependency ratio, and the young and elderly dependency ratios (Shakoor et al., 2021). In general, outcomes highlight a mature demographic structure in Eastern European countries, with low and declining fertility and relatively poor survival to old age (Botev, 2012). For this reason, a decrease in the population of working age leads to a corresponding increase of the elderly dependency rate (Borrescio-Higa & Valenzuela, 2021). On the one side, empirical studies highlight a future decrease in the elderly and, in turn, in the age dependency ratio, in established market economies, e.g. Europe, Japan and North America, performing an increasing number of old dependents (Lopreite & Mauro, 2017).

On the other hand, China and India seem to record an important demographic dividend (Golley & Tyers, 2012). A sharp increase in the age dependency ratio is expected as a consequence of the expanding share of elder population progressively leaving the labour market (Cleland & Machiyama, 2017; Kwan & Drolet, 2015; Okijie et al., 2023). Lastly, emerging economies report very young population due to high fertility and mortality rates (Alam et al., 2019; Higgins & Williamson, 1997; Verdugo, 2006). These countries will also experience future improvements in their economies (Bhattacharya & Haldar, 2015; Marois et al., 2021; Tabutin et al., 2020). Therefore, an increase of the old-age dependency ratio is to be expected not only in affluent countries, but also in emerging economies (Amanuma et al., 2023; Flenady et al., 2011; İnce Yenilmez, 2015).

Concluding Remarks

Measuring the age dependency ratio enables the design of adequate policies specifically addressed to youth or seniority, and could support the achievement of the SDGs concerning the working conditions also improved through a better education and gender equality policies.

References

Abbas, M., Alharthi, M., & Arshad, Z. (2024). Population dependency ratios and economic growth in SAARC nations: Insights on aging and youth dynamics. *Journal of Infrastructure, Policy and Development, 8*(8), 4875.

Abdullah, B., & Mansoor, K. (2024). Employment crisis and decent work deficits for youth in India. *Journal of Social and Economic Development, 26*(3), 734–758.

Aday, L. A. (1994). Health status of vulnerable populations. *Annual Review of Public Health, 15*(1), 487–509.

Admassie, A. (2002). Explaining the high incidence of child labour in Sub-Saharan Africa. *African Development Review, 14*(2), 251–275.

Ahmed, H. (2014). The socio-economic and political impacts of youth bulge: The case of Sudan. *Journal of Social Science Studies, 1*(2), 224–235.

Alam, M. M., Murad, M. W., Molla, R. I., Rahman, K. M., & Khondaker, T. R. (2019). Global population stabilisation policy and declining work-age population: A threat to global economic sustainability. *International Journal of Environment and Sustainable Development, 18*(4), 369–386.

Amanuma, N., Zusman, E., & Langlet, D. (2023). The relationship between female and younger legislative representation and performance on the Sustainable Development Goals (SDGs). *Environmental Research Letters, 18*(5), Article 054018.

An, C. B., & Jeon, S. H. (2006). Demographic change and economic growth: An inverted-U shape relationship. *Economics Letters, 92*(3), 447–454.

Andrews, D., & Cingano, F. (2014). Public policy and resource allocation: Evidence from firms in OECD countries. *Economic Policy, 29*(78), 253–296.

Ashraf, Q. H., Weil, D. N., & Wilde, J. (2013). The effect of fertility reduction on economic growth. *Population and Development Review, 39*(1), 97–130.

Barrientos, A., Gorman, M., & Heslop, A. (2003). Old age poverty in developing countries: Contributions and dependence in later life. *World Development, 31*(3), 555–570.

Beer, L., & Boswell, T. (2002). The resilience of dependency effects in explaining income inequality in the global economy: A cross-national analysis, 1975–1995. *Journal of World Systems Research, 8*(1), 30–59.

Bellettini, G., & Ceroni, C. B. (2000). Social security expenditure and economic growth: An empirical assessment. *Research in Economics, 54*(3), 249–275.

Bernheim, B. D. (1987). The economic effects of social security: Toward a reconciliation of theory and measurement. *Journal of Public Economics, 33*(3), 273–304.

Bhattacharya, G., & Haldar, S. K. (2015). Does demographic dividend yield economic dividend? India, a case study. *Economics Bulletin, 35*(2), 1274–1291.

Bidisha, S. H., Abdullah, S. M., & Islam, M. M. (2019). Is there any long run relationship between dependency ratio and economic growth? Evidences from Bangladesh. *Demography India, 48*(2), 105–120.

Bloom, D. E., Boersch-Supan, A., McGee, P., & Seike, A. (2011). Population aging: Facts, challenges, and responses. *Benefits and Compensation International, 41*(1), 22.

Borrescio-Higa, F., & Valenzuela, P. (2021). Does education mitigate the effect of population aging on health expenditure? A panel data study of Latin American Countries. *Journal of Aging and Health, 33*(7–8), 585–595.

Botev, N. (2012). Population ageing in Central and Eastern Europe and its demographic and social context. *European Journal of Ageing, 9*(1), 69–79.

Brown, P., Lauder, H., Ashton, D., Yingje, W., & Vincent-Lancrin, S. (2008). Education, globalisation and the future of the knowledge economy. *European Educational Research Journal, 7*(2), 131–156.

Chandrasekhar, C. P., Ghosh, J., & Roychowdhury, A. (2006). The 'demographic dividend' and young India's economic future. *Economic and Political Weekly*, 5055–5064.

Choudhry, M. T., Marelli, E., & Signorelli, M. (2016). Age dependency and labour productivity divergence. *Applied Economics, 48*(50), 4823–4845.

Cleland, J., & Machiyama, K. (2017). The challenges posed by demographic change in sub-Saharan Africa: A concise overview. *Population and Development Review, 43*, 264–286.

Cristea, M., Georgiana Noja, G., Dănăcică, D. E., & Ştefea, P. (2020). Population ageing, labour productivity and economic welfare in the European Union. *Economic Research-Ekonomska Istraživanja, 33*(1), 1354–1376.

Cruz, M., & Ahmed, S. A. (2018). On the impact of demographic change on economic growth and poverty. *World Development, 105*, 95–106.

D'Agostino, A., & Regoli, A. (2013). Life conditions and opportunities of young adults: Evidence from Italy in European comparative perspective. *Social Indicators Research, 113*(3), 1205–1235.

Da Silva, R. G. L., Chammas, R., & Novaes, H. M. D. (2021). Rethinking approaches of science, technology, and innovation in healthcare during the COVID-19 pandemic: The challenge of translating knowledge infrastructures to public needs. *Health Research Policy and Systems, 19*(1), 104.

Daly, M., & Lewis, J. (2000). The concept of social care and the analysis of contemporary welfare states. *The British Journal of Sociology, 51*(2), 281–298.

Dang, A., Arora, D., & Rane, P. (2020). Role of digital therapeutics and the changing future of healthcare. *Journal of Family Medicine and Primary Care, 9*(5), 2207–2213.

Dao, M. Q. (2012). Population and economic growth in developing countries. *International Journal of Academic Research in Business and Social Sciences, 2*(1), 6.

Darby, J., & Melitz, J. (2008). Social spending and automatic stabilizers in the OECD. *Economic Policy, 23*(56), 716–756.

Disney, R. (1996). Ageing and saving. *Fiscal Studies, 17*(2), 83–101.

Disney, R. (2004). Are contributions to public pension programmes a tax on employment? *Economic Policy, 19*(39), 268–311.

Egidi, G., Salvati, L., & Vinci, S. (2020). The long way to tipperary: City size and worldwide urban population trends, 1950–2030. *Sustainable Cities and Society, 60*, Article 102148.

Ehrlich, I., & Zhong, J. G. (1998). Social security and the real economy: An inquiry into some neglected issues. *The American Economic Review, 88*(2), 151–157.

Esping-Andersen, G. (2000). The sustainability of welfare states into the twenty-first century. *International Journal of Health Services, 30*(1), 1–12.

Ezeh, A. C., Bongaarts, J., & Mberu, B. (2012). Global population trends and policy options. *The Lancet, 380*(9837), 142–148.

Fehr, H., Jokisch, S., & Kotlikoff, L. J. (2008). Fertility, mortality and the developed world's demographic transition. *Journal of Policy Modeling, 30*(3), 455–473.

Feijóo, C., Kwon, Y., Bauer, J. M., Bohlin, E., Howell, B., Jain, R., & Xia, J. (2020). Harnessing artificial intelligence (AI) to increase wellbeing for all: The case for a new technology diplomacy. *Telecommunications Policy, 44*(6), Article 101988.

Flenady, V., Koopmans, L., Middleton, P., Frøen, J. F., Smith, G. C., Gibbons, K., & Ezzati, M. (2011). Major risk factors for stillbirth in high-income countries: A systematic review and meta-analysis. *The Lancet, 377*(9774), 1331–1340.

Frankowska, A., Głowacka-Toba, A., Rasińska, R., & Prussak, E. (2015). Students entering the labour market, their hopes, expectations and opportunities in the context of sustainable economic development. *Journal of International Studies, 7*(3), 209–222.

French, E., & Jones, J. (2012). Public pensions and labor supply over the life cycle. *International Tax and Public Finance, 19*(2), 268–287.

Galasso, V., & Profeta, P. (2004). Lessons for an ageing society: The political sustainability of social security systems. *Economic Policy, 19*(38), 64–115.

Gerland, P., Raftery, A. E., Ševčíková, H., Li, N., Gu, D., Spoorenberg, T., & Wilmoth, J. (2014). World population stabilization unlikely this century. *Science, 346*(6206), 234–237.

Golley, J., & Tyers, R. (2012). Demographic dividends, dependencies, and economic growth in China and India. *Asian Economic Papers, 11*(3), 1–26.

Gonzalez-Eiras, M., & Niepelt, D. (2012). Ageing, government budgets, retirement, and growth. *European Economic Review, 56*(1), 97–115.

Gough, I. (1996). Social assistance in southern Europe. *South European Society and Politics, 1*(1), 1–23.

Gough, I., Bradshaw, J., Ditch, J., Eardley, T., & Whiteford, P. (1997). Social assistance in OECD countries. *Journal of European Social Policy, 7*(1), 17–43.

Grant, A., & Benton, T. G. (2000). Elasticity analysis for density-dependent populations in stochastic environments. *Ecology, 81*(3), 680–693.

Hadley, C., Belachew, T., Lindstrom, D., & Tessema, F. (2011). The shape of things to come? Household dependency ratio and adolescent nutritional status in rural and urban Ethiopia. *American Journal of Physical Anthropology, 144*(4), 643–652.

Hammer, B., Prskawetz, A., & Freund, I. (2015). Production activities and economic dependency by age and gender in Europe: A cross-country comparison. *The Journal of the Economics of Ageing, 5*, 86–97.

Harper, S. (2014). Economic and social implications of aging societies. *Science, 346*(6209), 587–591.

Higgins, M., & Williamson, J. G. (1997). Age structure dynamics in Asia and dependence on foreign capital. *Population and Development Review*, 261–293.

Hock, H., & Weil, D. N. (2012). On the dynamics of the age structure, dependency, and consumption. *Journal of Population Economics, 25*(3), 1019–1043.

Holzmann, R. (2013). Global pension systems and their reform: Worldwide drivers, trends and challenges. *International Social Security Review, 66*(2), 1–29.

Ilan, Y. (2021). Improving global healthcare and reducing costs using second-generation artificial intelligence-based digital pills: A market disruptor. *International Journal of Environmental Research and Public Health, 18*(2), 811.

Inácio, P., Gomes, J. J., Airaksinen, M., & Cavaco, A. (2018). Exploring sociodemographic and economic factors that promote adverse drug reactions reporting by patients. *Health Policy, 122*(3), 263–268.

İnce Yenilmez, M. (2015). Economic and social consequences of population aging the dilemmas and opportunities in the twenty-first century. *Applied Research in Quality of Life, 10*(4), 735–752.

Inman, R. P. (1982). Public employee pensions and the local labor budget. *Journal of Public Economics, 19*(1), 49–71.

Iwański, R. (2024). Growth prospects for the silver economy in the market segment of residential care services provided to dependent elderly people. *Economics and Business Review, 10*(2), 165–186.

James, K. S. (2011). India's demographic change: Opportunities and challenges. *Science, 333*(6042), 576–580.

Jane Osareme, O., Muonde, M., Maduka, C. P., Olorunsogo, T. O., & Omotayo, O. (2024). Demographic shifts and healthcare: A review of aging populations and systemic challenges. *International Journal of Science and Research Archive, 11*, 383–395.

Jeetoo, J., & Jaunky, V. C. (2023). Divergence of healthcare expenditure in sub-Saharan Africa. *Studies in Economics and Econometrics, 47*(1), 75–105.

Jokinen, N., Janicki, M. P., Keller, S. M., McCallion, P., Force, L. T., & National task group on intellectual disabilities and dementia practices. (2013). Guidelines for structuring community care and supports for people with intellectual disabilities affected by dementia. *Journal of policy and practice in intellectual disabilities, 10*(1), 1–24.

Kasnauskienė, G., & Andriuškaitė, M. (2017). Economic implications of ageing Lithuanian population. *Organizations and Markets in Emerging Economies, 8*(1), 44–62.

Khan, H. T., Addo, K. M., & Findlay, H. (2024). Public health challenges and responses to the growing ageing populations. *Public Health Challenges, 3*(3), Article e213.

Kim, H., Park, Y., & Lee, H. (2013). Empirical dynamics of regional aged dependency ratio in South Korea. *Journal of Economic Theory and Econometrics, 24*(2), 163–192.

Kulik, C. T., Ryan, S., Harper, S., & George, G. (2014). Aging populations and management. *Academy of Management Journal, 57*(4), 929–935.

Kwan, C., & Drolet, J. (2015). Towards age-inclusive sustainable development goals: Exploring the potential role and contributions of community development. *Community Development Journal, 50*(4), 589–607.

Laurenza, E., Quintano, M., Schiavone, F., & Vrontis, D. (2018). The effect of digital technologies adoption in healthcare industry: A case based analysis. *Business Process Management Journal, 24*(5), 1124–1144.

Leal Filho, W., Kovaleva, M., Tsani, S., Țîrcă, D. M., Shiel, C., Dinis, M. A. P., & Tripathi, S. (2023). Promoting gender equality across the sustainable development goals. *Environment, Development and Sustainability, 25*(12), 14177–14198.

Lee, D., & Yoon, S. N. (2021). Application of artificial intelligence-based technologies in the healthcare industry: Opportunities and challenges. *International Journal of Environmental Research and Public Health, 18*(1), 271.

Lee, R. (2002). The demographic transition: Three centuries of fundamental change. *Journal of Economic Perspectives, 17*(4), 167–190.

Lee, R., & Mason, A. (2006). What is the demographic dividend? *Finance and Development, 43*(3), 16.

Leff, N. H. (1969). Dependency rates and savings rates. *The American Economic Review, 59*(5), 886–896.

Loichinger, E., & Skirbekk, V. (2016). International variation in ageing and economic dependency: A cohort perspective. *Comparative Population Studies–zeitschrift Für Bevölkerungswissenschaft, 41*(2), 121–144.

Lopreite, M., & Mauro, M. (2017). The effects of population ageing on health care expenditure: A Bayesian VAR analysis using data from Italy. *Health Policy, 121*(6), 663–674.

Ludwig, A., Schelkle, T., & Vogel, E. (2012). Demographic change, human capital and welfare. *Review of Economic Dynamics, 15*(1), 94–107.

Madhavan, S., Schatz, E., & Clark, B. (2009). Effect of HIV/AIDS-related mortality on household dependency ratios in rural South Africa, 2000–2005. *Population Studies, 63*(1), 37–51.

Maestas, N., & Zissimopoulos, J. (2010). How longer work lives ease the crunch of population aging. *Journal of Economic Perspectives, 24*(1), 139–160.

Marois, G., Gietel-Basten, S., & Lutz, W. (2021). China's low fertility may not hinder future prosperity. *Proceedings of the National Academy of Sciences, 118*(40), Article e2108900118.

Marois, G., Zhelenkova, E., & Ali, B. (2022). Labour force projections in India until 2060 and implications for the demographic dividend. *Social Indicators Research, 164*(1), 477–497.

Mbata, A. O., Soyege, O. S., Nwokedi, C. N., Tomoh, B. O., Mustapha, A. Y., Balogun, O. D., & Iguma, D. R. (2024). Preventative medicine and chronic disease management: Reducing healthcare costs and improving long-term public health. *International Journal of Multidisciplinary Research and Growth Evaluation, 5*(06), 1584–1600.

Mechanic, D., & Tanner, J. (2007). Vulnerable people, groups, and populations: Societal view. *Health Affairs, 26*(5), 1220–1230.

Mehta, B. S., & Awasthi, I. C. (2025). Indian Youth's Journey from Education to Decent Work. *India Studies in Business and Economics.*

Miles, D. (2023). Macroeconomic impacts of changes in life expectancy and fertility. *The Journal of the Economics of Ageing, 24*, Article 100425.

Minas, R. (2016). The concept of integrated services in different welfare states from a life course perspective. *International Social Security Review, 69*(3–4), 85–107.

Muszyńska, M. M., & Rau, R. (2012). The old-age healthy dependency ratio in Europe. *Journal of Population Ageing, 5*(3), 151–162.

Navaneetham, K., & Dharmalingam, A. (2012). A review of age structural transition and demographic dividend in South Asia: Opportunities and challenges. *Journal of Population Ageing, 5*(4), 281–298.

Okijie, S. R., Effiong, U. E., & Iriabije, A. O. (2023). Youth demographics and their role in sustainable development: Emphasis on the 2030 agenda. *European Journal of Theoretical and Applied Sciences, 1*(6), 89–104.

Oppenheimer, V. K. (1994). Women's rising employment and the future of the family in industrial societies. *Population and Development Review*, 293–342.

Paterson, D. H., Govindasamy, D., Vidmar, M., Cunningham, D. A., & Koval, J. J. (2004). Longitudinal study of determinants of dependence in an elderly population. *Journal of the American Geriatrics Society, 52*(10), 1632–1638.

Pecchenino, R. A., & Utendorf, K. R. (1999). Social security, social welfare and the aging population. *Journal of Population Economics, 12*(4), 607–623.

Pezzulo, C., Hornby, G. M., Sorichetta, A., Gaughan, A. E., Linard, C., Bird, T. J., & Tatem, A. J. (2017). Subnational mapping of population pyramids and dependency ratios in Africa and Asia. *Scientific Data, 4*(1), 1–15.

Pfau-Effinger, B. (2012). Analyses of welfare-state reform policies towards long-term senior care in a cross-European perspective. *European Journal of Ageing, 9*(2), 151–154.

Phurisikul, P., & Phuakkong, C. (2025). The elderly welfare management of local administrative government organizations in Thailand. *International Journal on Research and Development-A Management Review, 14*(1), 25–33.

Preston, S. H. (1984). Children and the elderly: Divergent paths for America's dependents. *Demography, 21*(4), 435–457.

Preston, S. H. (2007). The changing relation between mortality and level of economic development. *International Journal of Epidemiology, 36*(3), 484–490.

Prince, M. J., Wu, F., Guo, Y., Robledo, L. M. G., O'Donnell, M., Sullivan, R., & Yusuf, S. (2015). The burden of disease in older people and implications for health policy and practice. *The Lancet, 385*(9967), 549–562.

Prskawetz, A., & Sambt, J. (2014). Economic support ratios and the demographic dividend in Europe. *Demographic Research, 30*, 963–1010.

Rai, S. M., Brown, B. D., & Ruwanpura, K. N. (2019). SDG 8: Decent work and economic growth—A gendered analysis. *World Development, 113*, 368–380.

Ram, R. (1982). Dependency rates and aggregate savings: A new international cross-section study. *The American Economic Review, 72*(3), 537–544.

Razin, A., Sadka, E., & Swagel, P. (2002). The aging population and the size of the welfare state. *Journal of Political Economy, 110*(4), 900–918.

Salvati, L., & Carlucci, M. (2011). The economic and environmental performances of rural districts in Italy: Are competitiveness and sustainability compatible targets? *Ecological Economics, 70*(12), 2446–2453.

Salvati, L., & Carlucci, M. (2014). A composite index of sustainable development at the local scale: Italy as a case study. *Ecological Indicators, 43*, 162–171.

Salvati, L., & Zitti, M. (2008). Regional convergence of environmental variables: Empirical evidences from land degradation. *Ecological Economics, 68*(1–2), 162–168.

Salvati, L., & Zitti, M. (2009). Substitutability and weighting of ecological and economic indicators: Exploring the importance of various components of a synthetic index. *Ecological Economics, 68*(4), 1093–1099.

Santacreu, A. M. (2016). Long-run economic effects of changes in the age dependency ratio. *Economic Synopses, 17*, 1–2.

Sekhri, N., Feachem, R., & Ni, A. (2011). Public-private integrated partnerships demonstrate the potential to improve health care access, quality, and efficiency. *Health Affairs, 30*(8), 1498–1507.

Senbekov, M., Saliev, T., Bukeyeva, Z., Almabayeva, A., Zhanaliyeva, M., Aitenova, N., & Fakhradiyev, I. (2020). The recent progress and applications of digital technologies in healthcare: A review. *International Journal of Telemedicine and Applications, 2020*(1), 8830200.

Shabat, M. E. (2020). Early retirement incentive programs as a human resources restructuring strategy in public sector: Theoretical perspective. *Review of Economics and Political Science, 5*(1), 69–90.

Shakoor, U., Rashid, M., Baloch, A. A., Husnain, M. I. U., & Saboor, A. (2021). How aging population affects health care expenditures in Pakistan? A Bayesian VAR analysis. *Social Indicators Research, 153*(2), 585–607.

Singh, A., & Prakash, G. (2010). Public–private partnerships in health services delivery: A network organizations perspective. *Public Management Review, 12*(6), 829–856.

Skirbekk, V., Dieleman, J. L., Stonawski, M., Fejkiel, K., Tyrovolas, S., & Chang, A. Y. (2022). The health-adjusted dependency ratio as a new global measure of the burden of ageing: A population-based study. *The Lancet Healthy Longevity, 3*(5), e332–e338.

Slaveva, K. (2023). Study on the changes in the age dependency ratio and their effects on the economy. *Население, 41*(2), 83–105.

Suzman, R., Beard, J. R., Boerma, T., & Chatterji, S. (2015). Health in an ageing world—What do we know? *The Lancet, 385*(9967), 484–486.

Tabutin, D., Schoumaker, B., Coleman, H., Dutreuilh, C., Reeve, P., Tovey, J., & van Hoorn Alkema, B. (2020). The demography of Sub-Saharan Africa in the 21st century. *Population, 75*(2), 165–286.

Taneja, A. (2018). Investment in South Asia's human capital: Ensuring the 2 Qs for a demographic dividend: "Investing in people is investing in inclusive growth". *Educational Quest – An International Journal of Education and Applied Social Sciences*, *9*(2), 149–159.

Tatem, A. J., Campbell, J., Guerra-Arias, M., De Bernis, L., Moran, A., & Matthews, Z. (2014). Mapping for maternal and newborn health: The distributions of women of childbearing age, pregnancies and births. *International Journal of Health Geographics, 13*(1), 2.

Thinley, S. (2021). Health and care of an ageing population: Alignment of health and social systems to address the need. *Journal of Health Management, 23*(1), 109–118.

Tsai, I. J., Chu, C. C., & Chung, C. F. (2000). Demographic transition and household saving in Taiwan. *Population and Development Review*, 174–193.

Uhlenberg, P. (2009). Children in an aging society. *Journals of Gerontology Series b: Psychological Sciences and Social Sciences, 64*(4), 489–496.

Vardopoulos, I., Tsilika, E., Sarantakou, E., Zorpas, A. A., Salvati, L., & Tsartas, P. (2021). An integrated SWOT-PESTLE-AHP model assessing sustainability in adaptive reuse projects. *Applied Sciences, 11*(15), 7134.

Varsha Pramod, P., & Ramachandran, R. (2023). Youth employment for inclusive growth: A review and research agenda in global perspective with special reference to India. *Journal of Global Entrepreneurship Research, 13*(1), 9.

Verdugo, R. R. (2006). Workers, workers' productivity and the dependency ratio in Germany: Analysis with implications for social policy. *Population Research and Policy Review, 25*(5), 547–565.

Wang, L., & Wang, M. (2025). The impact of population aging on regional economic growth and adaptive reforms of the social security system. *Journal of Cross-Disciplinary Sustainability Innovations, 1*(1), 16–20.

Wei, Z., & Hao, R. (2010). Demographic structure and economic growth: Evidence from China. *Journal of Comparative Economics, 38*(4), 472–491.

Williams, J. S., Walker, R. J., & Egede, L. E. (2016). Achieving equity in an evolving healthcare system: Opportunities and challenges. *The American Journal of the Medical Sciences, 351*(1), 33–43.

Wilson, G. A., Kelly, C. L., Briassoulis, H., Ferrara, A., Quaranta, G., Salvia, R., Detsis, V., Curfs, M., Cerda, A., El-Aich, A., Liu, H., Kosmas, C., Alados, C. L., Imeson, A., Landgrebe-Trinkunaite, R., Salvati, L., Naumann, S., Danwen, H., Iosifides, T., … Zhang, P. (2017). Social memory and the resilience of communities affected by land degradation. *Land Degradation and Development, 28*(2), 383–400.

Wolff, J., Pauling, J., Keck, A., & Baumbach, J. (2020). The economic impact of artificial intelligence in health care: Systematic review. *Journal of Medical Internet Research, 22*(2), Article e16866.

Youssef, H., Elden, N. M. K., & Abou Ali, H. (2018). Realizing the prospects of the demographic dividend in selected Arab Countries. *Epidemiology, Biostatistics, and Public Health, 15*(3).

Zammitti, A., Moreno-Morilla, C., Romero-Rodríguez, S., Magnano, P., Marcionetti, J., & Parola, A. (2024). Health and well-being, quality education, gender equality, decent work and inequalities: The contribution of psychology in achieving the objectives of the Agenda 2030. *Frontiers in Psychology, 15*, 1353102.

Zhang, H., Zhang, H., & Zhang, J. (2015). Demographic age structure and economic development: Evidence from Chinese provinces. *Journal of Comparative Economics, 43*(1), 170–185.

Zhang, J., & Zhang, J. (2005). The effect of life expectancy on fertility, saving, schooling and economic growth: Theory and evidence. *Scandinavian Journal of Economics, 107*(1), 45–66.

Zhao, K. (2014). Social security and the rise in health spending. *Journal of Monetary Economics, 64*, 21–37.

Adolescent Fertility Rate

4

Introduction

Early marriage and fertility arise as major issues in social inequality and global health, due to their (short- and long-term) detrimental social and health effects on young women and their children (Gennari, 2013; Ghose & John, 2017; Saleh, 2022). However, the share of women aged 20–24 married (or in a union) before the age 15 sharply was found decreasing at the global scale (Singh & Darroch, 2000). In establishing causes and size of this decline, social factors and norms acquire a relevant role (Wilson et al., 2017). Increasing trends were observed in contexts promoting early marriage and childbearing while, decreasing trends characterize countries favouring delayed marriage, rather than births outside of marriage (Klepinger et al., 1999; Sedgh et al., 2015; Singh, 1998). A crucial factor sharply affecting marriage and childbearing is the level of education of girls (Fatusi & Hindin, 2010). To this purpose, empirical findings outline an indirect relationship between women level of education and the desired number of children (Chandra-Moul et al., 2013; Chigona & Chetty, 2007; Liang et al., 2019).

Effective family planning programmes would support women in choosing the timing and desired number of children (Blum et al., 2004; Chilisa, 2002; Temin et al., 2010). Sexual education should serve as a basis for informed decision-making (Görgen et al., 1993; Masuda & Yamauchi, 2020; Yang & Gaydos, 2010). Following the Sustainable Development Goals (SDGs) of United Nations, the adolescent fertility rate quantifies the live-births per 1000 individuals aged 10–14 years and 15–19 years (Sanhueza et al., 2023). According with Judiasih et al. (2020) and Cordova-Pozo et al. (2023), this indicator act as a key monitoring for SDG5 (gender equality) and SDG3 (good health and wellbeing).

Literature Review

When analysing the issue of adolescent fertility, factors such as work, education, health, and race, were considered in the literature (e.g. Furstenberg 1990; Gutelius, 1970; Tietze, 1977; McAnarney, 1978; Spellacy et al., 1978; Rhode, 1993; Markovitz et al., 2005; Penman-Aguilar et al., 2013). Human capital brought to the market by young women is as a major determinant of their earnings capacity (Rauscher, 2011; Santelli et al., 2000; Shanahan et al., 2002). From a societal perspective, this issue acquires a crucial relevance for the financial burden on taxpayers and individuals (Cobb-Clark et al., 2019; Fletcher & Wolfe, 2009; Melby et al., 2008). Despite these concerns, literature remarks that, in real life, these young women rely on a system of private and public sources of income, in addition to the young mothers' own earnings (Handler, 1987). These patterns of support affect the governmental propensity to improve existing policies and design

C. Ciaschini and L. Salvati, *Sustainable Development*, Sustainable Development Goals Series,
https://doi.org/10.1007/978-3-032-12559-0_4

new ones for the well-being of young mothers and their children. Government efforts should induce teenagers to delay pregnancies and births, thus reducing the need for assistance (Alemayehu et al., 2010).

Alternatively, policies should expand the income of young mothers in private jobs. An effective way to tackle the issue should be that of women empowerment, through assistance in completing school and acquiring job skills to better face the job market. Additional policies should improve the support of absent fathers, enabling them to pay for childcare. Given a panel of the National Research Council, Klepinger et al. (1999) concluded that teen mothers are more prone to social and economic disadvantage than those postponing childbearing after their twenties. Identifying childbearing as the primary cause of adverse outcomes is not an easy research task. Accordingly, these outcomes rather result from the disadvantaged backgrounds of teenage mothers and other differences between them and women avoiding early motherhood. In the latter case, public policies to reduce teenage pregnancy and childbearing would fail in improving the living conditions of disadvantaged young women (Barreto et al., 1992).

Assuming an inverse relationship between the age at first birth and the years of schooling (Waite & Moore, 1978), a general consensus exists on the negative effects of reproduction in adolescence on education (Card & Wise, 1978; Rindfuss et al., 1980). In general, women from advantaged backgrounds were more likely to graduate than those coming from disadvantaged contexts, and the likelihood to graduate reduced with their age of pregnancy and childbirth (Upchurch & McCarthy, 1990; Geronimus & Korenman, 1992; Kearney & Levine, 2012). Ribar (1999) reported negative disaggregated effects for white and Hispanics, and positive for blacks, on completed years of schooling (see also Ribar, 1994; Klepinger et al., 1995; Hotz et al., 1997, 2005). However, exceptions were also documented in some empirical studies (Bronars & Groggers, 1994; Hoffman et al., 1993; Olsen & Farkas, 1989) and the aggregated impact of family size, and economic well-being, should be investigated more carefully (Ashcraft et al., 2013; Lee, 2010; Yakusheva & Fletcher, 2015).

Considering health consequences, individual records and some official statistics seem biased towards complicated cases and wealthier socioeconomic classes able to pay hospital fees (Paccoud et al., 2021). Furthermore, in poor areas, a substantial proportion of, even severely complicated births, may not occur in hospitals and health centres (Cookson et al., 2016; Gordon et al., 2020; Victora et al., 2011). In many developing countries, there is a low level of completeness of vital registration and the classification of the cause of death is often inadequate (United Nations, 2019). In most developed countries, on the other hand, the quality of vital registration is satisfactory but teenage mothers are a rare occurrence. Consequently, the small, retrieved samples lead to unreliable estimates and prevalence rates of maternal mortality (Finlay et al., 2017). Research on the relationship between teenage fertility and maternal health outcomes also depends on the level of aggregation of age classes (Blanc et al., 2013; Harris et al., 2011; Nove et al., 2014).

Pregnancy complications are one of the most frequently consequences of teenage pregnancy and, in some countries, the leading cause of death among teenagers (United Nations, 2019). In Bangladesh, about 40 per cent of deaths among teenage mothers can be attributed to maternal causes (Islam et al., 2017), when compared with most developed countries where teenage fertility is low and complications of pregnancy and delivery usually account for less than 1 per cent of deaths (Sayem & Nury, 2011). Data for several countries also show that, apart from high parity women in the age group 40–49, there is a higher risk of maternal deaths among teenage girls than among older girls (Restrepo-Méndez & Victora, 2014). The causes of death among teenagers are often very different from those of older women (Lisonkova et al., 2017). A study of all health institutions in Oran (Algeria) for the time period encompassing 1971 and 1988 revealed that, in general, two-thirds of maternal deaths were from direct obstetric causes among which haemorrhage, sepsis and hypertensive disorders of pregnancy (Gage, 1999). However, women below twenty were more likely to suffer from hypertensive disorders of pregnancy.

Some studies reported that teenagers are more likely to die for abortion complications than older women (Ajayi et al., 1991). One study in Nigeria found that, in the period 1973–1985, induced abortion accounted for at least 70 per cent of all maternal deaths in the 15–19-year age group and for all deaths among girls aged 15 or less, compared to 8 per cent of all maternal deaths among women aged 20–24 (Unuigbe, 1988). Induced abortion is also a leading cause of maternal morbidity among teenagers in some developing countries (Kennedy et al., 2011; Miller, 1981; Temmerman, 2017). However, it is difficult to assess the prevalence of induced abortion and related complications. Retrospective surveys tend to produce gross underestimates, undoubtedly a reflection of the sensitive nature of the subject and, in some settings, the illegal or restricted nature of induced abortion (Wang & Qiao, 2020). However, a large number of clandestine abortions still occur, particularly in contexts where schools do not allow pregnant girls to continue their studies (Gupta and da Costa Leite, 1999; Lindberg et al., 2016; Abdul Rahim et al., 2024). The health consequences of improperly performed induced abortion can be severe, including post-abortive infertility, perforation, cervical injury and trauma (Jutte et al., 2010; Shaw et al., 2006; Yu et al., 2016). In addition, early sexual activity can carry other long-term health implication, like malignant tumours in the female reproductive system (Francesconi, 2008; Levine et al., 2007; Menacker et al., 2004).

Empirical Findings

This work relies on the adoption of official statistics based on provisional estimates, surveys and empirical models pertaining adolescent fertility and early birth rate (Dioikitopoulos & Varvarigos, 2023; Kane et al., 2013; Pantazis & Clark, 2018). Direct retrospective inquiries typically produce less accurate estimates depending on people's ability to recall an event occurred many months or years earlier (Fawole, 2008; Hill et al., 2007; Wood & Weinstein, 1988). They also depend on the existence of adequate survey or census programmes (Baffour et al., 2013; Peristera & Kostaki, 2007; Rafferty, 2013). The database adopted for the analysis was retrieved from the World Bank Indicator website (Abou Zahr et al. 2015; Kohno et al., 2020; Singh et al., 2025). Data available for the study cover 192 countries along 60 years between 1960 and 2020, offering comparable surveys with annual basis (Becedas et al., 2022; Durowaa-Boateng et al., 2023; Skinner, 2018). The indicator's mean value decreased from about 86 births per thousand women in 1960 to only 40 births in 2020. Periods with relevant decreases, as in the 1960s and the 1970s, were recorded over the whole time series. Similar trends were recorded between 2000 and 2010. Other periods were characterized by a slight and more stable decrease, as between 1984 and 1987 (Chavula et al., 2022; Pandey et al., 2023; Wilkins et al., 2022).

Considering the spatial distribution of this demographic indicator, almost all African countries displayed particularly high values for 1960. A high adolescent fertility rate persisted also in South America, in developing countries of South-East Asia and, partly, in Middle East countries. The spatial distribution of the same indicator for 2020 remained substantially unaltered, although the absolute values per country decreased moderately. Relevant variations also emerged in Middle Eastern countries and other major countries, such as India and China, shifting towards lower fertility at a young age.

Descriptive Statistics

Table 4.1 shows the global evolution of the adolescent fertility rate between 1960 and 2019 using summary statistics. On average, adolescent fertility rate assumed high values for 1960 (more than 100 births over 1000 adolescents). This value has progressively reduced throughout decades, reaching 45 in 2019. Even the median value, representing the central point of the distribution between all countries, has comparably decreased, moving from 103 to 37 within the same period. This generalized decrease indicates that maternity

Table 4.1 Descriptive statistics representing the distribution of adolescent fertility rate (births per 1000 women aged 15–19) across world countries by decade

Metric	1960	1970	1980	1990	2000	2010	2019
Mean	101.2	97.1	88.1	76.6	63.3	53.8	45.5
Minimum	4.4	4.3	4.1	2.3	1.1	0.4	0.3
Maximum	232.5	222.6	225.8	223.7	217.2	204.3	180.5
Norm. range	2.3	2.3	2.5	2.9	3.4	3.8	4.0
Median	103	93.0	80.0	69.3	50.8	44.0	37.5
25 percentile	48.4	48	42.45	32.8	24.43	15.9	12.05
75 percentile	145.9	139.8	122.3	106.0	94.9	81.3	67.8
Skewness	0.2	0.4	0.6	0.7	0.9	1.0	1.1
Kurtosis	– 1.0	– 0.9	– 0.6	– 0.2	0.3	0.6	0.7

Source Authors' elaboration of World Bank indicators

in adolescent age has become always less frequent in most of the countries worldwide.

Beside the decrease of average and median values, empirical data point out the reduction in maximum and minimum values. Some countries reported rates higher than 230 births over 1000 adolescents for 1960 and the maximum value decreased to about 180 for 2019. Contemporaneously, the minimum value has moved from 4.4 to only 0.3, highlighting that adolescent maternity has almost disappeared as a demographically relevant phenomenon in some advanced economies. Normalized range increases over time, from 2.25 in 1960 to nearly 4 in 2019, due to the decrease in the value of the mean value compared to the decrease in the maximum value.

Skewness and kurtosis indicate that differences among countries remained pronounced in 2019 compared with the beginning of the observed time series (Salvati & Carlucci, 2014). The increase over time in skewness (from 0.2 to 1.1) especially highlights that the statistical distribution of the adolescent fertility rate has become even more unbalanced across world regions (Salvati & Zitti, 2008). Today, most of the investigated countries have very low rates, but there are some contexts with still higher levels. Meanwhile, the increase in kurtosis suggests that the statistical distribution of the indicator has become more concentrated around central values (Salvati & Carlucci, 2011), but with the presence of extreme cases that prolong the tails (Salvati & Zitti, 2009). In summary, data confirm a generalized reduction of fertility in adolescent years—a positive signal attributable to different factors, as the improvement of female education, a wider access to reproductive healthcare services, and a higher awareness among young women. Nevertheless, strong disparities among countries still exist (sensu Egidi et al., 2020), and in some areas of the world adolescent maternity continues to represent a relevant challenge.

Considering summary values of the indicator computed by decade, the observed trend clearly reflects a deep change in reproductive behaviours and social, cultural and healthcare contexts, at least in many areas of the world. The continuous and significant decrease in female adolescent fertility worldwide, is the result of multiple factors including a higher access to education, particularly in the healthcare and social sectors, and generalized increases in the average age at first birth. In addition, increasingly widespread campaigns raising awareness to responsible reproductive behaviours, and a generalized cultural transformation involving almost all countries in the world, encouraged many girls to postpone maternity focussing on work and study (e.g. Vardopoulos et al., 2021). This portraits a positive evolution in terms of reproductive health and opportunities for young women. Nevertheless, the variability observed in the data, highlights the persistence of notable disparities between countries.

Correlation Analysis

Countries with accelerated dynamics, and showing high values of the indicator at the beginning of the observation time period (1960), resulted in a slower growth in the final year (2020). Countries such as the Republic of Gambia and Liberia experienced higher values for 1960 than for 2019. Countries such as Somalia and the Solomon Islands, have an accelerated growth in the final year, exhibiting even lower values in the initial years of the investigated time series. Figure 4.1 shows the evolution of two representative correlation indices throughout the investigated time interval. Considering both indices, the years performing particularly high correlation regimes, are those between 2001 and 2015, and indicate a similar development path across world countries. The largest differences between the two correlation indexes were observed for 1993 and 1997, suggesting an increasingly inhomogeneous demographic pattern across countries, possibly mitigated in the subsequent years.

Multidimensional Analysis

A Principal Component Analysis (PCA) considering the whole input data analysed above in this chapter extracted a first component (PC1) capturing most of the variability (90%) and possibly including the impact of broad socioeconomic patterns strongly associated with adolescent fertility (e.g. education levels, healthcare access, or cultural factors). A second component (PC2) accounted for 7.5% of the explained variance. Figure 4.2 provides a visual representation of the overall development path of adolescent fertility rates along the first two components (labelled as 'Dim1' and 'Dim2'). Most of 'country' points—primarily affected by PC1—were projected near the origin of the axes, whereas limited variability is shown with reference to PC2. Outliers, located far from the main clusters, represent countries with a peculiar path for the demographic indicator under scrutiny. For instance, Somalia (SOM) can be interpreted as a country with distinctive secondary trends in adolescent fertility rate, possibly related to factors of cultural, economic, or demographic nature.

Azerbaijan (AZE) is a country characterized by very low adolescent fertility rates and relevant socioeconomic aspects latently highlighted in the PCA biplot. Mozambique (MOZ) revealed as a country with significantly higher adolescent fertility rates, as it is intrinsically and primarily associated with PC1. Niger (NER), Chad (TCD), and Georgia (GEO), were also projected farther away from the central cluster, indicating variations in adolescent fertility rates possibly affected by local factors. A worth mentioning outlier country is Maldives (MDV), whose position in the biplot reflects a low adolescent fertility rate, likely connected with factors as an improved education and women healthcare conditions along with reduced early marriages.

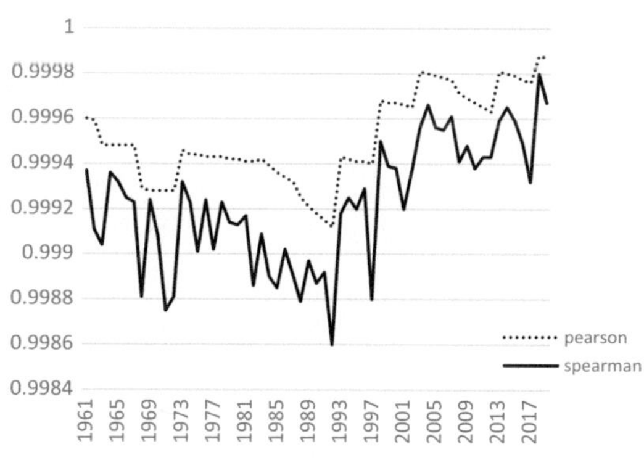

Fig. 4.1 Trends over time in parametric (Pearson) and non-parametric (Spearman) correlation coefficients within the (−1) lagged time series of adolescent fertility rate (births per 1000 women aged 15–19). *Source* Authors' elaboration of World Bank indicators

Discussion

Teenage pregnancy is a complex event, possibly influencing the harmonious development of personality (Lesthaeghe & Surkyn, 1988). For these reasons, support should derive not only from the partner (Chowdry et al., 2011), but also from the families of adolescent subjects and from external subjects as doctors, psychologists and counsellors (Berrington et al., 2016; Freake

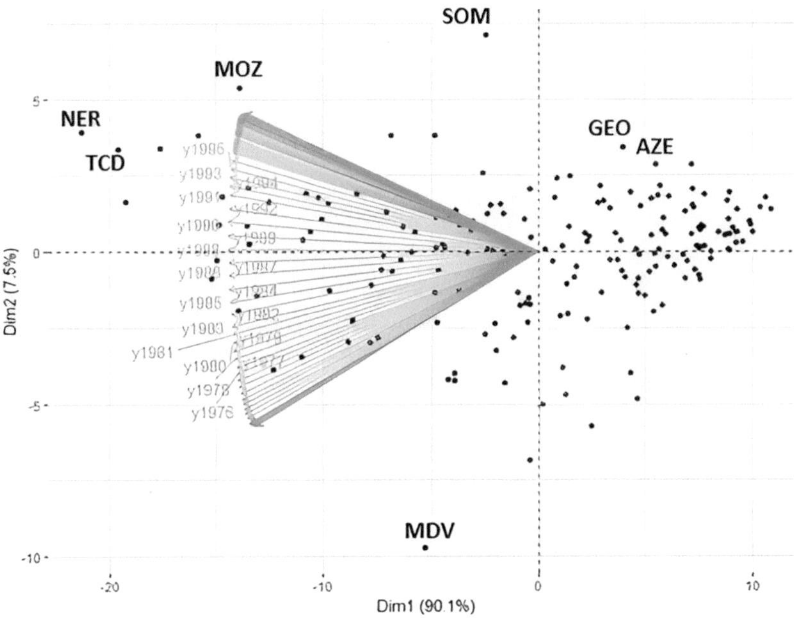

Fig. 4.2 Results of a Principal Component Analysis illustrated in a biplot reflecting the distribution of variables' (i.e. years) loadings and rows (i.e. countries) scores along the first two axes. *Source* Authors' elaboration of World Bank indicators

et al., 2007; Madarasova Geckova et al., 2010). Teenager parents often share similar factors (e.g. Bright et al., 2017), and are unfrequently capable to provide financial support for teenage mothers and their children—possibly complicating the relationship with the partner and the children (Furstenberg et al., 1987, 1990; Robert et al., 2005; Wellings et al., 1999). In this context, promoting programmes of sexual education and prevention that actively involve teenage girls is particularly effective in reducing the risk of early pregnancy (Barrington-Leigh & Escande, 2018; Parcel & Bixby, 2016; Philliber et al., 2002). In addition, access to adequate social and healthcare assistance services can make a significant difference in improving the living conditions of teenagers and their children (Gassman-Pines et al., 2015; Mechanic & Tanner, 2007; Nair et al., 2015). Lastly, it is crucial to strengthen the community and scholastic support networks to provide effective and continuous assistance to people living in conditions of vulnerability (Asma et al., 2020; Hák et al., 2016; Lafortune et al., 2020).

Concluding Remarks

Demographic indicators related to the attainment of Sustainable Development Goals–as previously mentioned in the introduction section—should be monitored, measured, and kept accurately updated. Further developments could concern the effectiveness of state aid and public prevention programmes. In addition, the net impact of public and private aids—from the partner and the parents' families to the neighbourhood—on the economic and social welfare of children, should be carefully investigated.

References

Abdul Rahim, K., Egglestone, N. J., Tsagareli, I. G., Usmani, W., Meherali, S., & Lassi, Z. S. (2024). Mental health outcomes beyond the post-partum period among adolescent mothers: A systematic review and meta-analysis. *Health Psychology and Behavioral Medicine, 12*(1), 2305741.

AbouZahr, C., De Savigny, D., Mikkelsen, L., Setel, P. W., Lozano, R., Nichols, E., & Lopez, A. D. (2015). Civil registration and vital statistics: Progress in the data revolution for counting and accountability. *The Lancet, 386*(10001), 1373–1385.

Ajayi, A. A., Marangu, L. T., Miller, J., & Paxman, J. M. (1991). Adolescent sexuality and fertility in Kenya: A survey of knowledge, perceptions, and practices. *Studies in Family Planning, 22*(4), 205–216.

Alemayehu, T., Haider, J., & Habte, D. (2010). Determinants of adolescent fertility in Ethiopia. *Ethiopian Journal of Health Development*, *24*(1).

Ashcraft, A., Fernández-Val, I., & Lang, K. (2013). The consequences of teenage childbearing: Consistent estimates when abortion makes miscarriage non-random. *The Economic Journal, 123*(571), 875–905.

Asma, S., Lozano, R., Chatterji, S., Swaminathan, S., de Fátima Marinho, M., Yamamoto, N., & Murray, C. J. (2020). Monitoring the health-related sustainable development goals: Lessons learned and recommendations for improved measurement. *The Lancet, 395*(10219), 240–246.

Baffour, B., King, T., & Valente, P. (2013). The modern census: Evolution, examples and evaluation. *International Statistical Review, 81*(3), 407–425.

Barreto, T., Campbell, O. M. R., Davies, J. L., Fauveau, V., Filippi, V. G. A., Graham, W. J., Mamdani, M., Rooney, C. I. F., & Toubia, N. F. (1992). Investigating induced abortion in developing countries: Methods and problems. *Studies in Family Planning, 23*(3), 159–170.

Barrington-Leigh, C., & Escande, A. (2018). Measuring progress and well-being: A comparative review of indicators. *Social Indicators Research, 135*(3), 893–925.

Becedas, R. R., Fruchtman, C. S., Dincu, I., De Savigny, D., & Muñoz, D. C. (2022). Addressing the evidence gap in the economic and social benefits of civil registration and vital statistics systems: A systematic review. *Public Health Reviews, 43*, 1604560.

Berrington, A., Roberts, S., & Tammes, P. (2016). Educational aspirations among UK Young Teenagers: Exploring the role of gender, class and ethnicity. *British Educational Research Journal, 42*(5), 729–755.

Blanc, A. K., Winfrey, W., & Ross, J. (2013). New findings for maternal mortality age patterns: Aggregated results for 38 countries. *PLoS ONE, 8*(4), Article e59864.

Blum, R. W., & Nelson-Mmari, K. (2004). The health of young people in a global context. *Journal of Adolescent Health, 35*(5), 402–418.

Bright, T., Felix, L., Kuper, H., & Polack, S. (2017). A systematic review of strategies to increase access to health services among children in low and middle income countries. *BMC Health Services Research, 17*(1), 252.

Bronars, S. G., & Grogger, J. (1994). The economic consequences of unwed motherhood: Using twin births as a natural experiment. *The American Economic Review, 84*(5), 1141–1156.

Card, J. J., & Wise, L. L. (1978). Teenage mothers and teenage fathers: The impact of early childbearing on the parents' personal and professional lives. *Family Planning Perspectives, 10*(4), 199–205.

Chandra-Mouli, V., Camacho, A. V., & Michaud, P. A. (2013). WHO guidelines on preventing early pregnancy and poor reproductive outcomes among adolescents in developing countries. *Journal of Adolescent Health, 52*(5), 517–522.

Chavula, M. P., Zulu, J. M., & Hurtig, A. K. (2022). Factors influencing the integration of comprehensive sexuality education into educational systems in low-and middle-income countries: A systematic review. *Reproductive Health, 19*(1), 196.

Chigona, A., & Chetty, R. (2007). Girls' education in South Africa: Special consideration to teen mothers as learners. *Journal of Education for International Development.*

Chilisa, B. (2002). National policies on pregnancy in education systems in Sub-Saharan Africa: The case of Botswana. *Gender and Education, 14*(1), 21–35.

Chowdry, H., Crawford, C., & Goodman, A. (2011). The role of attitudes and behaviours in explaining socio-economic differences in attainment at age 16. *Longitudinal and Life Course Studies, 2*(1), 59–76.

Cobb-Clark, D. A., Salamanca, N., & Zhu, A. (2019). Parenting style as an investment in human development. *Journal of Population Economics, 32*(4), 1315–1352.

Cookson, R., Propper, C., Asaria, M., & Raine, R. (2016). Socio-economic inequalities in health care in England. *Fiscal Studies, 37*(3–4), 371–403.

Cordova-Pozo, K. L., Anishettar, S. S., Kumar, M., & Chokhandre, P. K. (2023). Trends in child marriage, sexual violence, early sexual intercourse and the challenges for policy interventions to meet the sustainable development goals. *International Journal for Equity in Health, 22*(1), 250.

Dioikitopoulos, E., & Varvarigos, D. (2023). Delay in childbearing and the evolution of fertility rates. *Journal of Population Economics, 36*(3), 1545–1571.

Durowaa-Boateng, A., Yildiz, D., & Goujon, A. (2023). A Bayesian model for the reconstruction of education- and age-specific fertility rates. *Demographic Research, 49*, 809–848.

Egidi, G., Salvati, L., & Vinci, S. (2020). The long way to tipperary: City size and worldwide urban population trends, 1950–2030. *Sustainable Cities and Society, 60*, Article 102148.

Fatusi, A. O., & Hindin, M. J. (2010). Adolescents and youth in developing countries: Health and development issues in context. *Journal of Adolescence, 33*(4), 499–508.

Fawole, O. I. (2008). Economic violence to women and girls: Is it receiving the necessary attention? *Trauma, Violence, and Abuse, 9*(3), 167–177.

Finlay, J. E., Norton, M. K., & Mejía-Guevara, I. J. I. J. (2017). Adolescent fertility and child health: The interaction of maternal age, parity and birth intervals in determining child health outcomes. *International Journal of Child Health and Nutrition, 6*(1), 16–33.

Fletcher, J. M., & Wolfe, B. L. (2009). Education and labor market consequences of teenage childbearing:

Evidence using the timing of pregnancy outcomes and community fixed effects. *Journal of Human Resources, 44*(2), 303–325.

Francesconi, M. (2008). Adult outcomes for children of teenage mothers. *Scandinavian Journal of Economics, 110*(1), 93–117.

Freake, H., Barley, V., & Kent, G. (2007). Adolescents' views of helping professionals: A review of the literature. *Journal of Adolescence, 30*(4), 639–653.

Furstenberg Jr, F. F., Brooks-Gunn, J., & Morgan, S. P. (1987). Adolescent mothers and their children in later life. *Family Planning Perspectives*, 142–151.

Furstenberg Jr, F. F., Levine, J. A., & Brooks-Gunn, J. (1990). The children of teenage mothers: Patterns of early childbearing in two generations. *Family Planning Perspectives*, 54–61.

Gage, A. J. (1999). The social implications of adolescent fertility. *Population and Poverty in Developing Countries*, 120–143.

Gassman-Pines, A., Gibson-Davis, C. M., & Ananat, E. O. (2015). How economic downturns affect children's development: An interdisciplinary perspective on pathways of influence. *Child Development Perspectives, 9*(4), 233–238.

Gennari, P. J. (2013). adolescent pregnancy in developing countries. *International Journal of Childbirth Education, 28*(1).

Geronimus, A. T., & Korenman, S. (1992). The socioeconomic consequences of teen childbearing reconsidered. *The Quarterly Journal of Economics, 107*(4), 1187–1214.

Ghose, S., & John, L. B. (2017). Adolescent pregnancy: An overview. *International Journal of Reproduction, Contraception, Obstetrics and Gynecology, 6*(10), 4197–4203.

Gordon, T., Booysen, F., & Mbonigaba, J. (2020). Socioeconomic inequalities in the multiple dimensions of access to healthcare: The case of South Africa. *BMC Public Health, 20*(1), 289.

Görgen, R., Maier, B., & Diesfeld, H. J. (1993). Problems related to schoolgirl pregnancies in Burkina Faso. *Studies in Family Planning*, 283–294.

Gupta, N., & da Costa Leite, I. (1999). Adolescent fertility behavior: trends and determinants in Northeastern Brazil. *International Family Planning Perspectives*, 125–130.

Gutelius, M. F. (1970). Child-rearing attitudes of teen-age Negro girls. *American Journal of Public Health and the Nations Health, 60*(1), 93–104.

Hák, T., Janoušková, S., & Moldan, B. (2016). Sustainable development goals: A need for relevant indicators. *Ecological Indicators, 60*, 565–573.

Handler, J. F. (1987). The transformation of aid to families with dependent children: The family support act in historical context. *New York University Review of Law and Social Change, 16*, 457.

Harris, B., Goudge, J., Ataguba, J. E., McIntyre, D., Nxumalo, N., Jikwana, S., & Chersich, M. (2011). Inequities in access to health care in South Africa. *Journal of Public Health Policy, 32*(S1), S102–S123.

Hill, K., Lopez, A. D., Shibuya, K., & Jha, P. (2007). Interim measures for meeting needs for health sector data: Births, deaths, and causes of death. *The Lancet, 370*(9600), 1726–1735.

Hoffman, S. D., Foster, E. M., & Furstenberg, F. F. (1993). Reevaluating the costs of teenage childbearing. *Demography, 30*(1), 1–13.

Hotz, V. J., McElroy, S. W., & Sanders, S. G. (2005). Teenage childbearing and its life cycle consequences: Exploiting a natural experiment. *Journal of Human Resources, 40*(3), 683–715.

Hotz, V. J., Mullin, C. H., & Sanders, S. G. (1997). Bounding causal effects using data from a contaminated natural experiment: Analysing the effects of teenage childbearing. *The Review of Economic Studies, 64*(4), 575–603.

Islam, M. M., Islam, M. K., Hasan, M. S., & Hossain, M. B. (2017). Adolescent motherhood in Bangladesh: Trends and determinants. *PLoS ONE, 12*(11), Article e0188294.

Judiasih, S. D., Rubiati, B., Yuanitasari, D., Salim, E. F., & Safira, L. (2020). Efforts to eradicate child marriage practices in Indonesia: Towards sustainable development goals. *Journal of International Women's Studies, 21*(6), 135–149.

Jutte, D. P., Roos, N. P., Brownell, M. D., Briggs, G., MacWilliam, L., & Roos, L. L. (2010). The ripples of adolescent motherhood: Social, educational, and medical outcomes for children of teen and prior teen mothers. *Academic Pediatrics, 10*(5), 293–301.

Kane, J. B., Morgan, S. P., Harris, K. M., & Guilkey, D. K. (2013). The educational consequences of teen childbearing. *Demography, 50*(6), 2129–2150.

Kearney, M. S., & Levine, P. B. (2012). Why is the teen birth rate in the United States so high and why does it matter? *Journal of Economic Perspectives, 26*(2), 141–166.

Kennedy, E., Gray, N., Azzopardi, P., & Creati, M. (2011). Adolescent fertility and family planning in East Asia and the Pacific: A review of DHS reports. *Reproductive Health, 8*(1), 11.

Klepinger, D. H., Lundberg, S., & Plotnick, R. D. (1995). Adolescent fertility and the educational attainment of young women. *Family Planning Perspectives*, 23–28.

Klepinger, D., Lundberg, S., & Plotnick, R. (1999). How does adolescent fertility affect the human capital and wages of young women? *Journal of Human Resources*, 421–448.

Kohno, A., Techasrivichien, T., Suguimoto, S. P., Dahlui, M., Nik Farid, N. D., & Nakayama, T. (2020). Investigation of the key factors that influence the girls to enter into child marriage: A meta-synthesis of qualitative evidence. *PLoS ONE, 15*(7), Article e0235959.

Lafortune, G., Fuller, G., Schmidt-Traub, G., & Kroll, C. (2020). How is progress towards the sustainable development goals measured? Comparing four approaches for the EU. *Sustainability, 12*(18), 7675.

Lee, D. (2010). The early socioeconomic effects of teenage childbearing: A propensity score matching approach. *Demographic Research, 23*, 697–736.

Lesthaeghe, R., & Surkyn, J. (1988). Cultural dynamics and economic theories of fertility change. *Population and Development Review*, 1–45.

Levine, J. A., Emery, C. R., & Pollack, H. (2007). The well-being of children born to teen mothers. *Journal of Marriage and Family, 69*(1), 105–122.

Liang, M., Simelane, S., Fillo, G. F., Chalasani, S., Weny, K., Canelos, P. S., & Snow, R. (2019). The state of adolescent sexual and reproductive health. *Journal of Adolescent Health, 65*(6), S3–S15.

Lindberg, L., Santelli, J., & Desai, S. (2016). Understanding the decline in adolescent fertility in the United States, 2007–2012. *Journal of Adolescent Health, 59*(5), 577–583.

Lisonkova, S., Potts, J., Muraca, G. M., Razaz, N., Sabr, Y., Chan, W. S., & Kramer, M. S. (2017). Maternal age and severe maternal morbidity: A population-based retrospective cohort study. *PLoS Medicine, 14*(5), Article e1002307.

Madarasova Geckova, A., Tavel, P., van Dijk, J. P., Abel, T., & Reijneveld, S. A. (2010). Factors associated with educational aspirations among adolescents: Cues to counteract socioeconomic differences? *BMC Public Health, 10*(1), 154.

Markovitz, B. P., Cook, R., Flick, L. H., & Leet, T. L. (2005). Socioeconomic factors and adolescent pregnancy outcomes: Distinctions between neonatal and post-neonatal deaths? *BMC Public Health, 5*(1), 79.

Masuda, K., & Yamauchi, C. (2020). How does female education reduce adolescent pregnancy and improve child health? Evidence from Uganda's universal primary education for fully treated cohorts. *The Journal of Development Studies, 56*(1), 63–86.

McAnarney, E. R. (1978). Adolescent pregnancy—A national priority. *American Journal of Diseases of Children, 132*(2), 125–126.

Mechanic, D., & Tanner, J. (2007). Vulnerable people, groups, and populations: Societal view. *Health Affairs, 26*(5), 1220–1230.

Melby, J. N., Conger, R. D., Fang, S. A., Wickrama, K. A. S., & Conger, K. J. (2008). Adolescent family experiences and educational attainment during early adulthood. *Developmental Psychology, 44*(6), 1519.

Menacker, F., Martin, J. A., MacDorman, M. F., & Ventura, S. J. (2004). Births to 10–14 year-old mothers, 1990–2002: Trends and health outcomes. *National Vital Statistics Reports, 53*(7), 1–20.

Miller, S. H. (1981). Young adolescent mothers-how they and their children fare. *Early Child Development and Care, 7*(4), 265–278.

Nair, M., Baltag, V., Bose, K., Boschi-Pinto, C., Lambrechts, T., & Mathai, M. (2015). Improving the quality of health care services for adolescents, globally: A standards-driven approach. *Journal of Adolescent Health, 57*(3), 288–298.

Nove, A., Matthews, Z., Neal, S., & Camacho, A. V. (2014). Maternal mortality in adolescents compared with women of other ages: Evidence from 144 countries. *The Lancet Global Health, 2*(3), e155–e164.

Olsen, R. J., & Farkas, G. (1989). Endogenous covariates in duration models and the effect of adolescent childbirth on schooling. *Journal of Human Resources, 24*, 39–53.

Paccoud, I., Baumann, M., Le Bihan, E., Pétré, B., Breinbauer, M., Böhme, P., & Leist, A. K. (2021). Socioeconomic and behavioural factors associated with access to and use of personal health records. *BMC Medical Informatics and Decision Making, 21*(1), 18.

Pandey, S., Adair, T., & United Nations economic and social commission for Asia and the Pacific. (2023). Development of the civil registration and vital statistics system in Nepal: Making progress with civil registration, but lacking with vital statistics. *Asia-Pacific Sustainable Development Journal, 30*(2), 199–225.

Pantazis, A., & Clark, S. J. (2018). A parsimonious characterization of change in global age-specific and total fertility rates. *PLoS ONE, 13*(1), Article e0190574.

Parcel, T. L., & Bixby, M. S. (2016). The ties that bind: Social capital, families, and children's well-being. *Child Development Perspectives, 10*(2), 87–92.

Penman-Aguilar, A., Carter, M., Snead, M. C., & Kourtis, A. P. (2013). Socioeconomic disadvantage as a social determinant of teen childbearing in the US. *Public Health Reports, 128*(2_suppl1), 5–22.

Peristera, P., & Kostaki, A. (2007). Modeling fertility in modern populations. *Demographic Research, 16*, 141–194.

Philliber, S., Kaye, J. W., Herrling, S., & West, E. (2002). Preventing pregnancy and improving health care access among teenagers: An evaluation of the Children's Aid Society-Carrera Program. *Perspectives on Sexual and Reproductive Health*, 244–251.

Rafferty, Y. (2013). International dimensions of discrimination and violence against girls: A human rights perspective. *Journal of International Women's Studies, 14*(1), 1–23.

Rauscher, E. (2011). Producing adulthood: Adolescent employment, fertility, and the life course. *Social Science Research, 40*(2), 552–571.

Restrepo-Méndez, M. C., & Victora, C. G. (2014). Maternal mortality by age: Who is most at risk? *The Lancet Global Health, 2*(3), e120–e121.

Rhode, D. L. (1993). Adolescent pregnancy and public policy. *Political Science Quarterly, 108*(4), 635–669.

Ribar, D. C. (1994). Teenage fertility and high school completion. *The Review of Economics and Statistics*, 413–424.

Ribar, D. C. (1999). The socioeconomic consequences of young women's childbearing: Reconciling disparate evidence. *Journal of Population Economics, 12*(4), 547–565.

Rindfuss, R. R., Bumpass, L., & St. John, C. (1980). Education and fertility: Implications for the roles women occupy. *American Sociological Review*, 431–447.

Robert, K. W., Parris, T. M., & Leiserowitz, A. A. (2005). What is sustainable development? Goals, indicators, values, and practice. *Environment: Science and Policy for Sustainable Development*, *47*(3), 8–21.

Saleh, M. A. (2022). Outcomes of teenage pregnancy at Benghazi Medical Center 2019–2020. *International Journal of Science Academic Research, 3*(3), 358–3602.

Salvati, L., & Carlucci, M. (2011). The economic and environmental performances of rural districts in Italy: Are competitiveness and sustainability compatible targets? *Ecological Economics, 70*(12), 2446–2453.

Salvati, L., & Carlucci, M. (2014). A composite index of sustainable development at the local scale: Italy as a case study. *Ecological Indicators, 43*, 162–171.

Salvati, L., & Zitti, M. (2008). Regional convergence of environmental variables: Empirical evidences from land degradation. *Ecological Economics, 68*(1–2), 162–168.

Salvati, L., & Zitti, M. (2009). Substitutability and weighting of ecological and economic indicators: Exploring the importance of various components of a synthetic index. *Ecological Economics, 68*(4), 1093–1099.

Sanhueza, A., Costa, J. C., Mújica, O. J., Carvajal-Velez, L., Caffe, S., Victora, C., & Barros, A. J. (2023). Trends and inequities in adolescent childbearing in Latin American and Caribbean countries across generations and over time: A population-based study. *The Lancet Child & Adolescent Health, 7*(6), 392–404.

Santelli, J. S., Lowry, R., Brener, N. D., & Robin, L. (2000). The association of sexual behaviors with socioeconomic status, family structure, and race/ethnicity among US adolescents. *American Journal of Public Health, 90*(10), 1582.

Sayem, A. M., & Nury, A. T. M. (2011). Factors associated with teenage marital pregnancy among Bangladeshi women. *Reproductive Health, 8*(1), 16.

Sedgh, G., Finer, L. B., Bankole, A., Eilers, M. A., & Singh, S. (2015). Adolescent pregnancy, birth, and abortion rates across countries: Levels and recent trends. *Journal of Adolescent Health, 56*(2), 223–230.

Shanahan, M. J., Mortimer, J. T., & Kruger, H. (2002). Adolescence and adult work in the twenty-first century. *Journal of Research on Adolescence, 12*(1), 99–120.

Shaw, M., Lawlor, D. A., & Najman, J. M. (2006). Teenage children of teenage mothers: Psychological, behavioural and health outcomes from an Australian prospective longitudinal study. *Social Science and Medicine, 62*(10), 2526–2539.

Singh, S. (1998). Adolescent childbearing in developing countries: A global review. *Studies in Family Planning*, 117–136.

Singh, S., & Darroch, J. E. (2000). Adolescent pregnancy and childbearing: Levels and trends in developed countries. *Family Planning Perspectives*, 14–23.

Singh, H., Singh, R., Singh, A., Prashar, N., & Lachyan, A. S. (2025). Empowering tomorrow: The role of comprehensive sexual education in preventing unintended pregnancies and sexually transmitted infections. *Journal of Comprehensive Health, 13*(1), 35–39.

Skinner, C. (2018). Issues and challenges in census taking. *Annual Review of Statistics and Its Application, 5*(1), 49–63.

Spellacy, W. N., Mahan, C. S., & Cruz, A. C. (1978). The adolescent's first pregnancy: A controlled study. *Southern Medical Journal, 71*(7), 768–771.

Temin, M., Levine, R., & Stonesifer, S. (2010). Start with a girl: A new agenda for global health. *Issues in Science and Technology, 26*(3), 33–40.

Temmerman, M. (2017). Adolescent mothers: Too young to be neglected. *The Lancet Child and Adolescent Health, 1*(3), 164–166.

Tietze, C. (1977). Legal abortions in the United States: Rates and ratios by race and age, 1972–1974. *Family Planning Perspectives*, 12–15.

United Nations (2019). Global and regional data for Sustainable Development Goal indicators. [E/2019/68] Statistics Division, department of economic and social affairs. Statistical annex, New York.

Unuigbe, J. A., Oronsaye, A. U., & Orhue, A. A. E. (1988). Abortion-related morbidity and mortality in Benin City, Nigeria: 1973–1985. *International Journal of Gynecology and Obstetrics, 26*(3), 435–439.

Upchurch, D. M., & McCarthy, J. (1990). Adolescent childbearing and high school completion in the 1980s: Have things changed? *Family Planning Perspectives*, 199–202.

Vardopoulos, I., Tsilika, E., Sarantakou, E., Zorpas, A. A., Salvati, L., & Tsartas, P. (2021). An integrated SWOT-PESTLE-AHP model assessing sustainability in adaptive reuse projects. *Applied Sciences, 11*(15), 7134.

Victora, C. G., Aquino, E. M., Do Carmo Leal, M., Monteiro, C. A., Barros, F. C., & Szwarcwald, C. L. (2011). Maternal and child health in Brazil: progress and challenges. *The Lancet*, *377*(9780), 1863–1876.

Waite, L. J., & Moore, K. A. (1978). The impact of an early first birth on young women's educational attainment. *Social Forces, 56*(3), 845–865.

Wang, Y., & Qiao, J. (2020). Trends and social determinants of adolescent marriage and fertility in China. *The Lancet Global Health, 8*(7), e873–e874.

Wellings, K., Wadsworth, J., Johnson, A., Field, J., & Macdowall, W. (1999). Teenage fertility and life chances. *Reviews of Reproduction, 4*(3), 184–190.

Wilkins, N. J., Rasberry, C., Liddon, N., Szucs, L. E., Johns, M., Leonard, S., & Oglesby, H. (2022). Addressing HIV/sexually transmitted diseases and pregnancy prevention through schools: An approach for strengthening education, health services, and school environments that promote adolescent sexual health and well-being. *Journal of Adolescent Health, 70*(4), 540–549.

Wilson, G. A., Kelly, C. L., Briassoulis, H., Ferrara, A., Quaranta, G., Salvia, R., Detsis, V., Curfs, M., Cerda, A., El-Aich, A., Liu, H., Kosmas, C., Alados, C. L., Imeson, A., Landgrebe-Trinkunaite, R., Salvati, L., Naumann, S., Danwen, H., Iosifides, T., … Zhang, P. (2017). Social memory and the resilience of communities affected by land degradation. *Land Degradation and Development, 28*(2), 383–400.

Wood, J. W., & Weinstein, M. (1988). A model of age-specific fecundability. *Population Studies, 42*(1), 85–113.

Yakusheva, O., & Fletcher, J. (2015). Learning from teen childbearing experiences of close friends: Evidence using miscarriages as a natural experiment. *Review of Economics and Statistics, 97*(1), 29–43.

Yang, Z., & Gaydos, L. M. (2010). Reasons for and challenges of recent increases in teen birth rates: A study of family planning service policies and demographic changes at the state level. *Journal of Adolescent Health, 46*(6), 517–524.

Yu, S. H., Mason, J., Crum, J., Cappa, C., & Hotchkiss, D. R. (2016). Differential effects of young maternal age on child growth. *Global Health Action, 9*(1), 31171.

Life Expectancy at Birth

5

Introduction

Life expectancy at birth measures how long a new born would live under unchanged conditions at the time of birth (Arriaga, 1984). The indicator practically provides the average expected life duration of a new born in the different countries worldwide (Kinsella, 1992), thus providing a global framework on health and welfare of a given population (Ara et al., 2022). Nevertheless, it is worth noting that this average value may diverge across countries, being affected by specific economic, demographic and social contexts (Oh et al., 2022). The duration of life, indeed, not only depends on health conditions, but also on the level of economic development, on the accessibility of healthcare services, education and life conditions, as well as other social and cultural aspects. The present study relies on a historical series of data on the indicator 'life expectancy at birth', retrieved from World Bank database, and covering a relatively long time period between 1960 and 2019.

By analysing these data, it is possible to note how throughout more than half a century there have been significant changes in its average value worldwide. Throughout time, indeed, there has been an almost linear growth of this value, reflecting progress in healthcare assistance systems, life conditions and social policies adopted by several countries (Buitrago & Caraballo, 2022). Nevertheless, this growth has not been uniform: some countries experienced fast and effective improvements (Johnson & Papageorgiou, 2020), while others have shown slower progresses or even stagnations (Baldwin & Huber, 2010). A relevant aspect to be considered in these studies is the distinction between genders, since factors affecting life expectancy differ between men and women. For this reason, the present analysis focuses specifically on the female life expectancy at birth (Austad, 2006).

In many contexts, women tend to live longer than men, thanks to biological, behavioural and social factors (Eskes & Haanen, 2007). Nevertheless, other variables influence this difference, as the access to care, life conditions, the social role, and the environment in which they develop (Langer et al., 2015). Besides the social and economic conditions (Karimi & Brazier, 2016), among the different factors affecting women life expectancy, the physical structure and the health status at birth should be considered (Setyadi et al., 2023). Moreover, health conditions and maternal nutrition, as well as environmental conditions during pregnancy, can also affect future longevity of females (Mimi et al., 2024). In this perspective, our work develops a preliminary statistical analysis, describing metrics as the average, maximum and minimum and their changes over the whole time period. This allows an overall vision on the general trend of female life expectancy worldwide. In addition, a bivariate analysis compares the values of the variable in the first and last year of analysis (i.e. 1960 and 2019), highlighting

C. Ciaschini and L. Salvati, *Sustainable Development*, Sustainable Development Goals Series,
https://doi.org/10.1007/978-3-032-12559-0_5

the differences and evolutions between these two years. A Principal Component Analysis (PCA) finally identifies the latent dimensions explaining the variations between the investigated variables.

The relevance of the demographic indicator adopted here consolidates in the context of Sustainable Development Goals (SDG) of 2030 UN Agenda. In particular, its connection with SDG3 (on health and welfare) highlights how life expectancy represents a crucial indicator to evaluate any progress in this direction (Rod et al., 2020). To this purpose, SDG 3.1 focuses specifically on the reduction of the avoidable mortality among new born, children under 5 years of age, and mothers, promoting actions to increase longevity and improve the life quality of future generations (Bartniczak et al., 2024). This direct connection between the specific indicator and the sustainable development goals highlights the importance of investments in healthcare (Hák et al., 2016), as well as social and economic policies improving life conditions, reducing inequalities and promoting a more equitable development worldwide (Nunes et al., 2016).

Literature Review

Life expectancy is a key metric assessing population health, and quantifying mortality along the entire lifetime, rather than focussing on mortality at a young age, as the infant and child mortality indices do (Smith & Pell, 2002). This led to a notably disproportionate distribution of healthcare services across the world. High—quality healthcare services become, then, a prerogative of rich countries while lacking in poor environments. Over the last two centuries, people achieved impressive progress in healthcare leading to further increases in life expectancy (Majer et al., 2011). Before pandemics, life expectancy at birth generally increased almost monotonously in most countries. In addition, improvements in life expectancy among high-income countries depend on gains made in old age ($\geq$ 65 years) in recent decades. A persisting heterogeneity emerged since the early 2010s.

While some Eastern European and Baltic countries experienced significant increases in life expectancy over the past decade, others perform slower improvements and, sometimes, stalls or even temporary reversals (McKinley et al., 2014).

For instance, life expectancy in the United States, England, Wales and Scotland, showed only limited increases over the past decade (Mackenbach et al., 2015). These atypical trends are connected with a lower level of mortality in old age and increased mortality in working age (Leon, 2011). Pandemics indirectly affected mortality due to other causes. Emerging evidence has highlighted the impact of delayed or lacking treatments for cancer or cardiovascular disease resulting in increased mortality, while the lockdown reduced the number of deaths due to accidents. Anyway, the female life expectancy at birth was found higher than the male one (Chen et al., 2007). Since the historical series under scrutiny considers information until 2019, the present study neglects the impact of Covid-19 pandemic (e.g. Aburto et al., 2022).

Economic and Demographic Issues

Life expectancy offers a complete framework of health conditions and social inequalities (Kunze, 2014; Laranjeira & Szrek, 2016; Palloni et al., 2009). Nevertheless, this measure is introduced within economic and policy contexts affecting the longevity levels of a society (Azomahou et al., 2009). The increasing life expectancy throughout time was related with the economic development of countries (Miles, 2023). In general, countries with a high level of economic development tend to have higher life expectancy than the developing ones. This because a stronger economy allows to highly invest in effective healthcare systems, prevention infrastructures, education and social policies, thus reducing healthcare inequalities (Bloom & Canning, 2003; Ferry & Vironen, 2011; Raphael & Bryant, 2006).

An emblematic example of this relationship was observed in European countries, where public policies aimed at ensuring a universal access to quality healthcare services, promoting healthy lifestyles and supporting prevention programmes (Garry & Checchi, 2020). These interventions contributed to increase the life expectancy, bringing it to more than 80 years in many developed countries. Conversely, in low-income countries, the lack of resources restrains the access to basic cares, fosters inequalities (Lutz & Striessnig, 2015) and favours unhealthy life conditions (Uchendu, 2018). The scarcity of healthcare infrastructure, the poor nutrition, the absence of prevention policies and the presence of wars and political instability are factors reducing life expectancy (Hannum & Buchmann, 2005).

Public policies play a crucial role in determining differences in life expectancy between populations (Li et al., 2025). Programmes of public healthcare oriented to prevention, vaccinations, screening and health education, are crucial to reduce chronic and infective diseases affecting mortality (Teh et al., 2025). In addition, social policies dealing with income inequalities and access to resources contribute to the improvement of life conditions of more vulnerable classes and favour increases in longevity (Hansen & Lønstrup, 2015). Welfare policies and social security measures ensure an economic support to older people and families in need (e.g. Wilson et al., 2017), reducing the disparities in healthcare connected to poverty (Fearon, 2003; He & Li, 2020; Lakioti et al., 2025). The promotion of healthy life environments, with attention to pollution, liveability and nutrition education, are further aspects to be faced to improve life quality and, in turn, life expectancy (Erdoğan et al., 2019).

The increase in life expectancy also implies relevant economic implications (Graham & Power, 2004). On the one side, it means an extension of the female productive period (Crampton, 2011); on the other side, it implies challenges connected to sustainability of welfare systems and healthcare assistance (McLeman, 2010). Demographic growth of old population requires adaptation policies, as the increase of resources destined to healthcare, social security and formation of a more inclusive job market for older people (Acemoglu & Johnson, 2007). In addition, the increasing longevity can generate pressures on retirement systems and on public resources, requiring structural reforms and investments policies in prevention and healthcare maintenance (Fan & Bo, 2025; Gianfredi et al., 2025; Haar & Rubenstein, 2012).

Policies of active ageing, promoting social participation, physical activity and continuous education, are effective tools to face these challenges, contributing to the maintenance of the life quality and reduce long-term healthcare costs (Golini, 2001). Generally speaking, life expectancy is not only a healthcare measure, but also a key indicator for economic conditions and public policies of a country (Allel et al., 2021). Investing in effective healthcare systems, inclusive social policies and prevention programmes represents the most effective strategy to improve the longevity of population (Luy & Gast, 2014) and ensure a sustainable development throughout time (Manderbacka et al., 2013).

Empirical Findings

In the last decades, geographic inequalities in life expectancy at birth persist, even with some improvements. In 1960, countries with the highest life expectancy (> 75–80 years) are mainly located in the most developed areas of the world. These included Western and Northern Europe (such as France, Scandinavia, United Kingdom), North America (United States and Canada) Oceania (Australia and New Zealand) and Japan. Conversely, many regions characterized by low life expectations, between 40 and 50 years, are located in the sub-Saharan Africa, in some parts of Southern Asia, as Afghanistan and Yemen, and in some regions of Southern America. Especially the decades immediately following World War II reflected a period of economic boom in Western Europe, North America and Japan, bringing significant improvements in life conditions, nutrition hygiene and health care. At the same time, many African countries were reaching the independence, with weak healthcare structures, civil wars and poverty contributing to low levels of life expectancy. Vaccinations and initiatives of global public healthcare against infective diseases were introduced only in the 1960s and mostly in wealthiest countries.

A more complex spatial framework emerged for 2019 and most of the world countries saw a substantial increase in life expectancy. In many African countries, life expectancy increased from less than 50 years to more than 60–65 years, even if, in some areas of Sahel and central Africa, critical conditions still persist. In Asia, countries as India, Bangladesh and Vietnam, have succeeded in going beyond 70 years, thanks to improvements in vaccinations, basic medical cares and health education. Latin American countries approached the levels of Europe and North America in terms of longevity. In 1960, developed countries (Europe, north America, Japan and Australia) were practically the only ones with life expectations over 70–75 years. In 2019, these countries maintained the leadership, with expectations often over 80 years even if other countries such as Chile, China, South Korea, and Turkey, are reaching similar levels. Nevertheless, significant inequalities in some areas marked by conflicts, poverty, political instability and health crises as HIV/AIDS (for instance, Africa and the Middle East), keep life expectations relegated between 55 and 65 years.

In both affluent and emerging economies, gains in life expectancy depends on multiple factors, including the global spread of vaccinations and antibiotics, a higher access to primary cares, the spread of female education and birth control. These aspects correlated with further improvements in nutrition regimes, access to drinking water, and hygienic conditions. International collaborations with organizations such as WHO (World Health Organization), UNICEF, and various Non-Governmental Organizations (NGOs), finally played a crucial role in reducing inequalities, contributing to bring life expectancy at birth (global average) from 52 years (1960) to 73 years (2019). In this framework, even the most disadvantaged areas, as sub-Saharan Africa (displaying life expectancies lower than 45 years in 1960), notably improved – even not exceeding 60 years. Even if extreme inequalities have been drastically reduced, they still represent one of the main challenges for developmental policies, highlighting that—despite some measurable progresses—the issue of equity in healthcare still remains unaddressed (e.g. Vardopoulos et al., 2021).

Descriptive Statistics

Confirming previous findings, life expectancy at birth increased continuously throughout the observed time period. The minimum value was observed in 1960, increasing until the maximum value performed in 2019. This increase reflects significant improvements in healthcare systems and the life condition of populations continuously gained over time. It is worth noting how public policies and healthcare innovations would have contributed to this steady growth. Moreover, the increase in life expectancy also highlights a general improvement in lifestyles and socioeconomic conditions. Despite a positive trend, some time periods showed more constant, or slightly decreasing, trends, suggesting possible future challenges.

Table 5.1 reports some descriptive statistics from global information extracted from World Bank database. Within this context, countries with missing data have been excluded. The minimum value of life expectancy, retrieved in 1960, equals 29 years, reaching 55.5 years in 2019. A negligible decline occurred between the 1970s and the 1990s. In 1960, the maximum value was 75.9 years, moving to 88.1 years in 2019. The statistical distribution displays a rather symmetric trend, since the difference between mean and median values was around 3 points throughout the whole period. The maximum value of skewness equals −0.15 in 1960, further reaching the minimum value (−0.77) in 1990, then increasing again until the value of −0.66 in 2019. Kurtosis assumes the minimum value (−1.24) in 1960, reaching the maximum in 1990 (−0.36), then increasing again, and finally decreasing to −0.37 in 2019.

The mean-whisker plot (Fig. 5.1) highlights important fluctuations over time. In particular, the general trend is stable and moderately increasing: differently from the typical, net increase throughout time, we observed a periodic fluctuation of the average value. The bars

Table 5.1 Descriptive statistics representing the distribution of historical series of female life expectancy at birth across world countries, 1960–2019

Metric	1960	1970	1980	1990	2000	2010	2019
Mean	55.93	60.23	64.21	67.35	69.22	72.58	75.16
Minimum	29	33.3	30.5	34.9	40	48.1	55.5
Maximum	75.9	77.3	80.4	81.9	84.6	86.3	88.1
Normal. range	0.84	0.73	0.78	0.7	0.64	0.53	0.43
Median	57.5	63.2	67.3	70.8	72.8	75.4	77.05
25 percentile	44.98	50.75	55.83	59.15	60.63	65.88	69.45
75 percentile	67.18	71.25	73.88	75.4	77.15	79.53	80.8
Skewness	− 0.15	− 0.38	− 0.64	− 0.77	− 0.74	− 0.75	− 0.66
Kurtosis	− 1.24	− 1.11	− 0.55	− 0.36	− 0.64	− 0.37	− 0.37

Source Authors' elaboration of World Bank indicators

show years with high variability reflecting significant differences between regions, countries and social groups. This makes the mean values less representative, and highlights a marked heterogeneity between data, possibly affected by recurrent socioeconomic factors, such as regional health crises, wars, and the (positive or negative) impact of political reforms.

The shape of the curve, often displaying values under the average—at least in some years—likely reflects periods of healthcare crisis. This cyclic variability could be the result of deep disparities internal to countries. For instance, in the poorer districts of England and Wales, women live in good health only for about 65% of their life, increasing to 81% in the most affluent areas. Factors as economy, access to cares, stress and care responsibilities, affect more and more women, contributing to the amplification of spatial inequalities. Women suffered more the effects of pandemics from psychological, nutritional and socioeconomic viewpoints. They often act as caregivers and, during recent lockdowns, they experienced an increase in domestic burden, worse mental health conditions, and a decrease in the quality of nutrition.

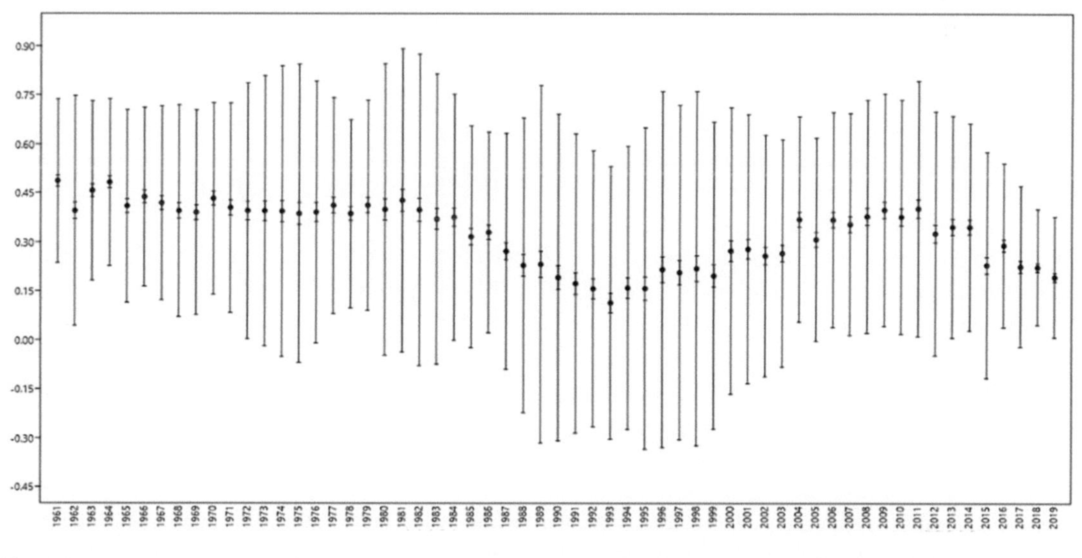

Fig. 5.1 A graphical summary of the statistical distribution of female life expectancy at birth across world countries based on a mean and whisker plot (standardized data with standard deviation and standard error). *Source* Authors' elaboration of World Bank indicators

Additionally, there are gender differences also connected with biological and behavioural factors. In general, women live longer than men, with an advantage of 5 years in wealthier countries. During pandemics, this difference has further exacerbated. In the United States, for instance, the gap widened to 6 years in 2021. Wide error bars in the plot reflect this heterogeneity, possibly deriving from biological factors, and negative behaviours as smoking. Many women face also chronic diseases as obesity, diabetes, or cardiovascular problems, preventing healthcare improvements and decreasing life expectancy, especially in Eastern countries. These factors are combined with other emerging risks and could have contributed to short term fluctuations in life expectancy. The considerable variability between years finally indicates the presence of shocks as pandemics, economic crisis or structural disparities.

More recent trends suggest a slowdown in the improvement of life expectancy and underline that differences in gender and social conditions are more evident in relation with the life quality in good health. Shifting from 1960 to 2019, the average age of female life expectancy at birth has increased sharply. Specifically, in 1960 the average age was around 56, while in 2019 it was 75. The whiskers, lines extending over and under the average, suggest a wide disparity among the different countries. In some countries, the female life expectancy was rather high, while in others it was very low, likely due to factors as poverty, conflicts, restrained access to healthcare services, high maternal mortality, and scarce hygiene. This notable increase reflects decades of global improvements in healthcare, public infrastructures, education (particularly for women and girls), and socioeconomic conditions at large. The narrowest interval of whiskers in 2019 indicates a reduced difference between countries—always more countries were aligned on higher levels of life expectancy. It is an encouraging signal to reduce inequalities, suggesting how even low-income or formerly disadvantaged countries have done significant progresses.

Based on earlier findings, it is questionable if increases in the global average life expectancy correspond to equally distributed improvements. It is a crucial outcome in terms of global healthcare equity, since it implies that an always lower number of women remains back simply due to the birthplace. In summary, these findings delineate a global positive trend: women in the word live longer, and this higher longevity is becoming more equally distributed between countries. It highlights the success of policy efforts addressing regional development, progresses in medical and technological fields, and the increasing attention to healthcare and women rights. The study finally highlights that female life expectancy, particularly in terms of welfare and life quality, is currently affected by an articulated set of health, economic, behavioural and structural factors, making this reality more complex compared with the past.

Correlation Analysis

Figure 5.2 highlights trends over time in Pearson (parametric) and Spearman (non-parametric) correlation indices exploring possible breaks in time series. On average, Spearman coefficients outperformed Pearson coefficients for almost the first ten years of observation. In the central years of the observation period, values of the two indices were rather similar, even with some peaks emerging for Pearson coefficients. In the last years, two almost overlapping trends were observed, with Spearman coefficients systematically below Pearson coefficients.

Multidimensional Analysis

Considering the outcomes of a Principal Component Analysis (PCA) that explores the evolution over time in life expectancy figures at the country scale, the overall variability in the investigated data seems to be well summarized in the first two components (PC1 and PC2). In particular, PC1 accounted for 92% of the total variance, making it the most significant dimension of dataset's variability. This suggests that PC1 likely captures the primary factor driving differences across world regions and countries, related to female Life Expectancy at Birth (LEB). Other

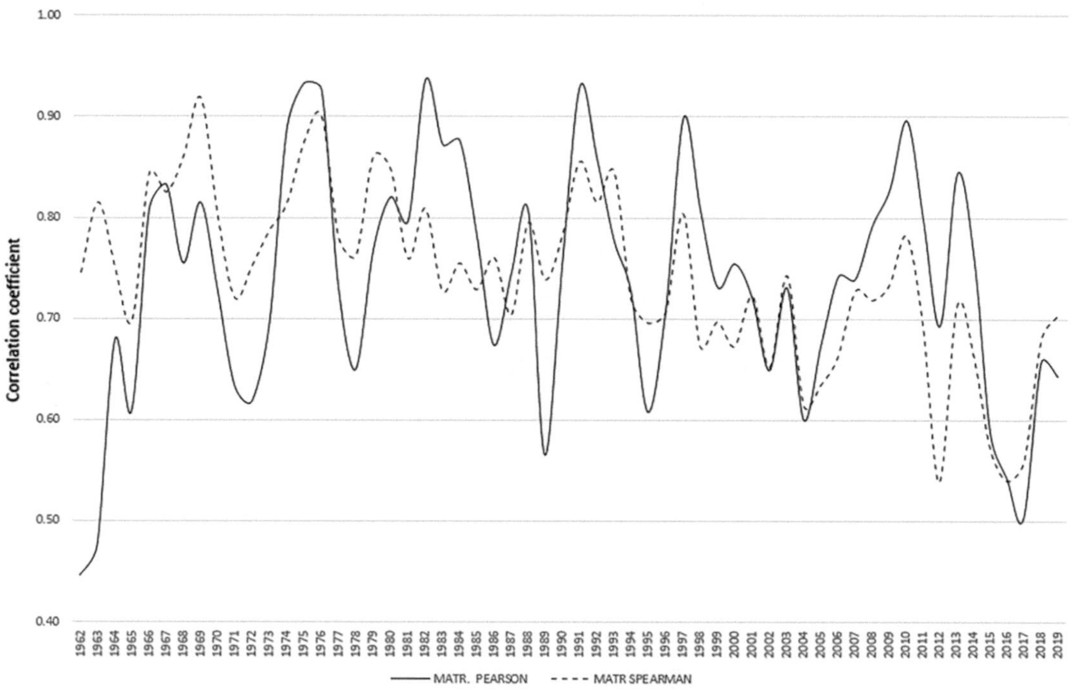

Fig. 5.2 Trends over time in parametric (Pearson—solid line) and non-parametric (Spearman—dotted line) correlation coefficients within the (−1) lagged time series of the average trend of female life expectancy at birth across world countries, 1960–2019. *Source* Authors' elaboration of World Bank indicators

variables (e.g. healthcare expenditure, per capita gross domestic product, or socioeconomic conditions) closely correlated with LEB might also contribute to PC1. PC2 explains 6% of the total variance and identifies a secondary source of variation, reflecting minor differences in socioeconomic conditions, environmental factors, regional disparities, or healthcare system characteristics (e.g. disease burdens) having a smaller (but still notable) impact on life expectancy. From PC3 onward, the contribution to the variance becomes negligible (< 1%).

Figure 5.3 depicts the PCA biplot derived from the first two components; points represent the analysed countries and the arrows reveal temporal dynamics (years) in life expectancy and the related factors. Countries in the positive side of Component 1 likely experienced health and socioeconomic improvements, while those stagnating along Component 2 perform secondary trends, likely related to outliers or anomalies (e.g. Maldives, Lesotho and Zimbabwe). Countries tightly clustered around the origin of the two axes likely experienced minimal change over time, suggesting relative stability in LEB—generally close to the global average of all analysed countries. Countries far from the origin suggest significant differences in magnitude and direction of change. Emerging economies (e.g. in Asia or Africa) might show strongly positive movements along Component 1. Economically fragile states or those experiencing prolonged crises (e.g. Zimbabwe) or climate change (e.g. Maldives), might show slower or negative trends. Maldives, located far on the negative side of Component 1 and slightly negative on Component 2, is a country with significantly low LEB due to adverse socioeconomic and environmental conditions as climate vulnerability or limited healthcare resources. Lesotho and Zimbabwe lay on the positive side of Component 1, while representing an extreme outlier with a very high Component 2 value. These countries shared several health challenges, as exceptionally high rates of HIV/AIDS and political instability affect their healthcare systems.

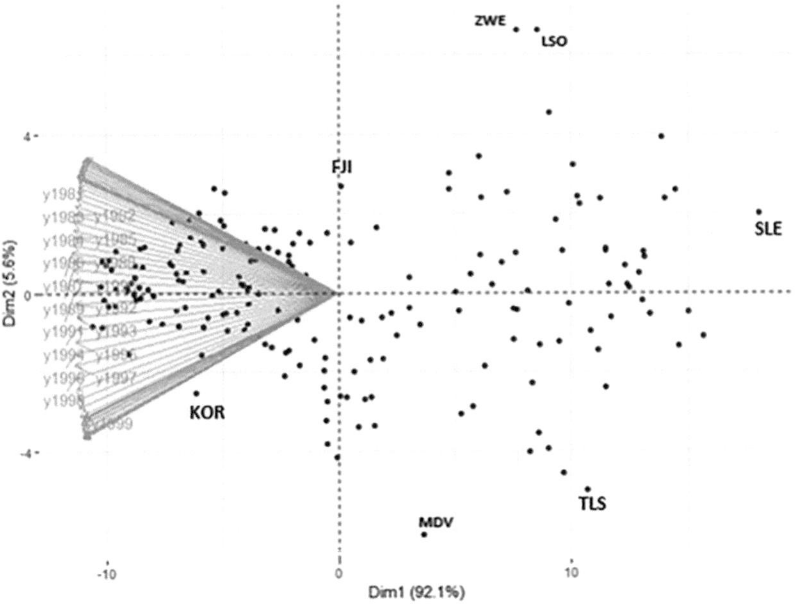

Fig. 5.3 Results of a principal component analysis illustrated in a biplot that reflects the distribution of variables' (i.e. years) loadings and rows (i.e. countries) scores along the first two axes. *Source* Authors' elaboration of World Bank indicators

The case of other countries, such as Sierra Leone, Fiji, South Korea and Timor Leste, is worth mentioning. Sierra Leone, in the top-right quadrant of the biplot, likely exhibits low life expectancy for women at birth, suggesting relevant issues as high maternal mortality, limited access to healthcare, and lower development indicators. South Korea's position strongly correlates with high female life expectancy, being in turn characterized by the highest overall quality of life at global level, an excellent healthcare system, and education. Fiji is located at the centre of the plot, slightly above the horizontal axis. It is characterized by a moderate or average level of life expectancy for women at birth compared with the other countries. Finally, Timor-Leste's position indicates lower life expectancy for women at birth, although potentially due to challenges like post-conflict recovery, limited healthcare infrastructure, and socioeconomic constraints impacting women's health and longevity.

Discussion

The intrinsic relevance of life expectancy at birth estimates has been recognized at the global level, encouraging its inclusion as key indicator within the 17 Sustainable Development Goals of the 2030 United Nations (UN) Agenda (Norheim et al., 2015). This work analyses the life expectancy of women in 192 countries of the world (Annandale, 2003) since, following a supposed 'paradox' ('women are sicker, but men die quicker') they lived longer than men, even if in worse health conditions (Halisçelik & Soytas, 2019). The empirical results of statistical analysis highlight a progressive increase in female life expectancy rate at birth over time (Wilmoth et al., 2000). Specifically, the indicator increased sharply in affluent economies, while a moderate growth was observed in disadvantaged countries, as Africa and several backward countries of Asia (Low et al., 2013). This finding could be also connected with traditions, cultures and habits of these countries, relegating women in subordinate positions with respect to men (Glei & Horiuchi, 2007). Within this context, women and girls suffer from practices relevantly shortening their life expectancy, also during pregnancy (Bledsoe et al., 1998).

Throughout the years, a better situation was highlighted in some affluent countries, achieving continuous medical progress, and thus experiencing higher life expectancy rates. Converging rates were frequently observed for countries with similar background contexts (Salvati & Zitti, 2008), possibly depending on (i) urbanization dynamics and the related lifestyle improvements (e.g. Egidi et al., 2020) and (ii) rural development (Salvati & Carlucci, 2011), especially in emerging economies. In particular, in European countries, as well as in some states of America, but also in China, life expectancy was about 85 years in the last years (Marphatia et al., 2017). In African countries, it oscillates between 60 and 70 years, with the Central African Republic showing a life expectancy at birth of 55.5 years (Aksan & Chakraborty, 2023).

Within the analysed period, tragic events, as wars, occurred in many countries have led to many deaths (Aburto & Van Raalte, 2018). For instance, the explosion of Chernobyl nuclear power plant in 1986 caused not only many deaths (Bard et al., 1997), but also diseases shortening the lives of many people in the years ahead (Ilyin et al., 1990). Additionally, AIDS negatively affected the life expectancy rate at birth, particularly for women (Edwards & Tuljapurkar, 2005). Nowadays, the recently introduced treatments and preventive measures have strongly reduced its spread and consequences (Congdon, 2006). These trends seem to be connected with a lower level of mortality in old age and increased mortality in working age (Milligan & Wise, 2015). A continuous monitoring of demographic indicators, including life expectancy at birth, and possibly using both descriptive statistics and analytical tools such as exploratory techniques (Salvati & Zitti, 2009) seems to be an appropriate tool accounting for sustainable development advancements in would regions and countries (Salvati & Carlucci, 2014).

Concluding Remarks

In a context of heterogeneous trajectories in life expectancy over recent decades, the high mortality induced by Covid–19 pandemic (not specifically considered in this study), should imply further challenges to population health. It specifically requires a more effective commitment of governments and policy makers to tackle global health issues.

References

Aburto, J. M., Schöley, J., Kashnitsky, I., Zhang, L., Rahal, C., Missov, T. I., & Kashyap, R. (2022). Quantifying impacts of the COVID-19 pandemic through life-expectancy losses: A population-level study of 29 countries. *International Journal of Epidemiology, 51*(1), 63–74.

Aburto, J. M., & Van Raalte, A. (2018). Lifespan dispersion in times of life expectancy fluctuation: The case of Central and Eastern Europe. *Demography, 55*(6), 2071–2096.

Acemoglu, D., & Johnson, S. (2007). Disease and development: The effect of life expectancy on economic growth. *Journal of Political Economy, 115*(6), 925–985.

Aksan, A. M., & Chakraborty, S. (2023). Life expectancy across countries: Convergence, divergence and fluctuations. *World Development, 168*, Article 106263.

Allel, K., Salustri, F., Haghparast-Bidgoli, H., & Kiadaliri, A. (2021). The contributions of public health policies and healthcare quality to gender gap and country differences in life expectancy in the UK. *Population Health Metrics, 19*, 1–11.

Annandale, E. (2003). Gender and Health. *International Journal of Health Services, 33*(3), 569–579.

Ara, I., Maqbool, M., & Gani, I. (2022). Reproductive health of women: Implications and attributes. *International Journal of Current Research in Physiology and Pharmacology*, 8–18.

Arriaga, E. E. (1984). Measuring and explaining the change in life expectancies. *Demography, 21*(1), 83–96.

Austad, S. N. (2006). Why women live longer than men: Sex differences in longevity. *Gender Medicine, 3*(2), 79–92.

Azomahou, T. T., Boucekkine, R., & Diene, B. (2009). A closer look at the relationship between life expectancy and economic growth. *International Journal of Economic Theory, 5*(2), 201–244.

Baldwin, K., & Huber, J. D. (2010). Economic versus cultural differences: Forms of ethnic diversity and public goods provision. *American Political Science Review, 104*(4), 644–662.

Bard, D., Verger, P., & Hubert, P. (1997). Chernobyl, 10 years after: Health consequences. *Epidemiologic Reviews, 19*(2), 187–204.

Bartniczak, B., Płachciak, A., Raszkowski, A., & Lewis, G.J. (2024). Good health and well-being: An assessment of sustainable development goal (SDG) No. 3 in the Sahel countries. *Sustainability, 16*(5), 2109.

Bledsoe, C., Banja, F., & Hill, A. G. (1998). Reproductive mishaps and Western contraception: An African challenge to fertility theory. *Population and Development Review*, 15–57.

Bloom, D., & Canning, D. (2003). The health and poverty of nations: From theory to practice. *Journal of Human Development, 4*(1), 47–71.

Buitrago, E. M., & Caraballo, M. Á. (2022). Measuring social diversity in economic literature: An overview for cross-country studies. *Journal of Economic Surveys, 36*(4), 880–934.

Chen, X. K., Wen, S. W., Fleming, N., Demissie, K., Rhoads, G. G., & Walker, M. (2007). Teenage pregnancy and adverse birth outcomes: A large population based retrospective cohort study. *International Journal of Epidemiology, 36*(2), 368–373.

Congdon, P. (2006). A model for geographical variation in health and total life expectancy. *Demographic Research, 14*, 157–178.

Crampton, A. (2011). Population aging and social work practice with older adults: Demographic and policy challenges. *International Social Work, 54*(3), 313–329.

Edwards, R. D., & Tuljapurkar, S. (2005). Inequality in life spans and a new perspective on mortality convergence across industrialized countries. *Population and Development Review, 31*(4), 645–674.

Egidi, G., Salvati, L., & Vinci, S. (2020). The long way to tipperary: City size and worldwide urban population trends, 1950–2030. *Sustainable Cities and Society, 60*, Article 102148.

Erdoğan, S., Yıldırım, D. Ç., & Gedikli, A. (2019). The relationship between CO_2 emissions and health indicators: The case of Turkey. *Economics Letters, 6*(1), 28–39.

Eskes, T., & Haanen, C. (2007). Why do women live longer than men? *European Journal of Obstetrics and Gynecology and Reproductive Biology, 133*(2), 126–133.

Fan, C., & Bo, L. (2025). Determinants of life expectancy in China: A comparative analysis of economic and non-economic factors. *African Journal of Reproductive Health, 29*(3), 160–167.

Fearon, J. D. (2003). Ethnic and cultural diversity by country. *Journal of Economic Growth, 8*, 195–222.

Ferry, M., & Vironen, H. (2011). Dealing with demographic change: Regional policy responses. *Geopolitics, History and International Relations, 3*(1), 25.

Garry, S., & Checchi, F. (2020). Armed conflict and public health: Into the 21st century. *Journal of Public Health, 42*(3), e287–e298.

Gianfredi, V., Nucci, D., Pennisi, F., Maggi, S., Veronese, N., & Soysal, P. (2025). Aging, longevity, and healthy aging: The public health approach. *Aging Clinical and Experimental Research, 37*(1), 1–12.

Glei, D. A., & Horiuchi, S. (2007). The narrowing sex differential in life expectancy in high-income populations: Effects of differences in the age pattern of mortality. *Population Studies, 61*(2), 141–159.

Golini, A. (2001). Demographic trends and population policies. *Futures, 33*(1), 27–41.

Graham, H., & Power, C. (2004). Childhood disadvantage and health inequalities: A framework for policy based on lifecourse research. *Child: Care, Health and Development, 30*(6), 671–678.

Haar, R. J., & Rubenstein, L. S. (2012). Health in fragile and post-conflict states: A review of current understanding and challenges ahead. *Medicine, Conflict and Survival, 28*(4), 289–316.

Hák, T., Janoušková, S., & Moldan, B. (2016). Sustainable Development Goals: A need for relevant indicators. *Ecological Indicators, 60*, 565–573.

Halisçelik, E., & Soytas, M. A. (2019). Sustainable development from millennium 2015 to sustainable development goals 2030. *Sustainable Development, 27*(4), 545–572.

Hannum, E., & Buchmann, C. (2005). Global educational expansion and socio-economic development: An assessment of findings from the social sciences. *World Development, 33*(3), 333–354.

Hansen, C. W., & Lønstrup, L. (2015). The rise in life expectancy and economic growth in the 20th century. *The Economic Journal, 125*(584), 838–852.

He, L., & Li, N. (2020). The linkages between life expectancy and economic growth: Some new evidence. *Empirical Economics, 58*(5), 2381–2402.

Ilyin, L. A., Balonov, M. I., Buldakov, L. A., Bur'Yak, V. N., Gordeev, K. I., Dement'Ev, S. I., & Tsyb, A. F. (1990). Radio contamination patterns and possible health consequences of the accident at the Chernobyl nuclear power station. *Journal of Radiological Protection, 10*(1), 3.

Johnson, P., & Papageorgiou, C. (2020). What remains of cross-country convergence? *Journal of Economic Literature, 58*(1), 129–175.

Karimi, M., & Brazier, J. (2016). Health, health-related quality of life, and quality of life: What is the difference? *Pharmaco Economics, 34*, 645–649.

Kinsella, K. G. (1992). Changes in life expectancy 1900–1990. *The American Journal of Clinical Nutrition, 55*(6), 1196S-1202S.

Kunze, L. (2014). Life expectancy and economic growth. *Journal of Macroeconomics, 39*, 54–65.

Lakioti, E., Pagonis, N., Flegkas, D., Itziou, A., Moustakas, K., & Karayannis, V. (2025). Social factors and policies promoting good health and well-being as a sustainable development goal: Current achievements and future pathways. *Sustainability, 17*(11), 5063.

Langer, A., Meleis, A., Knaul, F. M., Atun, R., Aran, M., Arreola-Ornelas, H., & Frenk, J. (2015). Women and health: The key for sustainable development. *The Lancet, 386*(9999), 1165–1210.

Laranjeira, E., & Szrek, H. (2016). Going beyond life expectancy in assessments of health systems' performance: Life expectancy adjusted by perceived health status. *International Journal of Health Economics and Management, 16*, 133–161.

Leon, D. A. (2011). Trends in European life expectancy: A salutary view. *International Journal of Epidemiology, 40*(2), 271–277.

Li, R., Zhan, Y., Tao, Y., Wang, R., Gu, X., He, Z., & Zhu, B. (2025). Economic and social determinants of life expectancy in China. *Frontiers in Public Health, 13*, 1554336.

Low, B. S., Parker, N., Hazel, A., & Welch, K. B. (2013). Life expectancy, fertility, and women's lives: A life-history perspective. *Cross-Cultural Research, 47*(2), 198–225.

Lutz, W., & Striessnig, E. (2015). Demographic aspects of climate change mitigation and adaptation. *Population Studies, 69*(supl), S69–S76.

Luy, M., & Gast, K. (2014). Do women live longer or do men die earlier? Reflections on the causes of sex differences in life expectancy. *Gerontology, 60*(2), 143–153.

Mackenbach, J. P., Karanikolos, M., Lopez Bernal, J., & Mckee, M. (2015). Why did life expectancy in Central and Eastern Europe suddenly improve in the 1990s? An analysis by cause of death. *Scandinavian Journal of Public Health, 43*(8), 796–801.

Majer, I. M., Nusselder, W. J., Mackenbach, J. P., & Kunst, A. E. (2011). Socioeconomic inequalities in life and health expectancies around official retirement age in 10 Western-European countries. *Journal of Epidemiology and Community Health, 65*(11), 972–979.

Manderbacka, K., Peltonen, R., Lumme, S., Keskimäki, I., Tarkiainen, L., & Martikainen, P. (2013). The contribution of health policy and care to income differences in life expectancy—A register based cohort study. *BMC Public Health, 13*, 1–9.

Marphatia, A. A., Ambale, G. S., & Reid, A. M. (2017). Women's marriage age matters for public health: A review of the broader health and social implications in South Asia. *Frontiers in Public Health, 5*, 269.

McKinley, W., Latham, S., & Braun, M. (2014). Organizational decline and innovation: Turnarounds and downward spirals. *Academy of Management Review, 39*(1), 88–110.

McLeman, R. (2010). Impacts of population change on vulnerability and the capacity to adapt to climate change and variability: A typology based on lessons from "a hard country." *Population and Environment, 31*, 286–316.

Miles, D. (2023). Macroeconomic impacts of changes in life expectancy and fertility. *The Journal of the Economics of Ageing, 24*, Article 100425.

Milligan, K., & Wise, D. A. (2015). Health and work at older ages: Using mortality to assess the capacity to work across countries. *Journal of Population Ageing, 8*, 27–50.

Mimi, M. B., Kibria, M. G., & Selim, M. M. I. (2024). How do economic, health, environmental and demographic factors affect life expectancy? A novel attempt for developed and developing economies. *International Journal of Sustainable Development and World Ecology, 31*(7), 776–791.

Norheim, O. F., Jha, P., Admasu, K., Godal, T., Hum, R. J., Kruk, M. E., & Peto, R. (2015). Avoiding 40% of the premature deaths in each country, 2010–30: Review of national mortality trends to help quantify the UN sustainable development goal for health. *The Lancet, 385*(9964), 239–252.

Nunes, A. R., Lee, K., & O'Riordan, T. (2016). The importance of an integrating framework for achieving the sustainable development goals: The example of health and well-being. *BMJ Global Health, 1*(3), Article e000068.

Oh, J., Min, J., Kang, C., Kim, E., Lee, J. P., Kim, H., & Lee, W. (2022). Excess mortality and the COVID-19 pandemic: Causes of death and social inequalities. *BMC Public Health, 22*(1), 2293.

Palloni, A., Milesi, C., White, R. G., & Turner, A. (2009). Early childhood health, reproduction of economic inequalities and the persistence of health and mortality differentials. *Social Science and Medicine, 68*(9), 1574–1582.

Raphael, D., & Bryant, T. (2006). The state's role in promoting population health: Public health concerns in Canada, USA, UK, and Sweden. *Health Policy, 78*(1), 39–55.

Rod, N. H., Bengtsson, J., Budtz-Jørgensen, E., Clipet-Jensen, C., Taylor-Robinson, D., Andersen, A. M. N., & Rieckmann, A. (2020). Trajectories of childhood adversity and mortality in early adulthood: A population-based cohort study. *The Lancet, 396*(10249), 489–497.

Salvati, L., & Carlucci, M. (2011). The economic and environmental performances of rural districts in Italy: Are competitiveness and sustainability compatible targets? *Ecological Economics, 70*(12), 2446–2453.

Salvati, L., & Carlucci, M. (2014). A composite index of sustainable development at the local scale: Italy as a case study. *Ecological Indicators, 43*, 162–171.

Salvati, L., & Zitti, M. (2008). Regional convergence of environmental variables: Empirical evidences from land degradation. *Ecological Economics, 68*(1–2), 162–168.

Salvati, L., & Zitti, M. (2009). Substitutability and weighting of ecological and economic indicators: Exploring the importance of various components of a synthetic index. *Ecological Economics, 68*(4), 1093–1099.

Setyadi, S., Kustanto, A., & Widiastuti, A. (2023). Life expectancy in Indonesia: The role of health infrastructure, political, and socioeconomic status. *Iranian Economic Review, 27*(3), 965–1005.

Smith, G. C., & Pell, J. P. (2002). Teenage pregnancy and risk of adverse perinatal outcomes associated with first and second births: Population based retrospective cohort study. *Obstetrical and Gynecological Survey, 57*(3), 136–137.

Teh, W. S., Ling, G. L. H. T., Rusli, N., Teh, H. H., Tan, K. P., Faisal, S., & Mohamed, M. S. A. (2025). Factors influencing life expectancy: Global evidence from 125 countries. *International Journal of Built Environment and Sustainability, 12*(2), 17–28.

Uchendu, F. N. (2018). Hunger influenced life expectancy in war-torn Sub-Saharan African countries. *Journal of Health, Population and Nutrition, 37*, 1–4.

Vardopoulos, I., Tsilika, E., Sarantakou, E., Zorpas, A. A., Salvati, L., & Tsartas, P. (2021). An integrated SWOT-PESTLE-AHP model assessing sustainability in adaptive reuse projects. *Applied Sciences, 11*(15), 7134.

Wilmoth, J. R. (2000). Demography of longevity: Past, present, and future trends. *Experimental Gerontology, 35*(9–10), 1111–1129.

Wilson, G. A., Kelly, C. L., Briassoulis, H., Ferrara, A., Quaranta, G., Salvia, R., Detsis, V., Curfs, M., Cerda, A., El-Aich, A., Liu, H., Kosmas, C., Alados, C. L., Imeson, A., Landgrebe-Trinkunaite, R., Salvati, L., Naumann, S., Danwen, H., Iosifides, T., … Zhang, P. (2017). Social memory and the resilience of communities affected by land degradation. *Land Degradation and Development, 28*(2), 383–400.

6 Carbon Dioxide Emissions

Introduction

In the last decades, climate change emerged as a key issue challenging governments and international institutions (Maâlej & Cabagnols, 2020). The trade-off between nature conservation and economic growth still fosters the debate in all global contexts also for the relevance they have in the fight of climate change (Salvati & Carlucci, 2014; Vardopoulos et al., 2021; Barbier & Burgess, 2022). These two elements may reflect the unbalanced geography of physical capital, since developing countries own plenty of natural resources, while developed countries have limited endowments (Reilly et al., 2002; Radhi, 2009; Pecl et al., 2017). Natural resources, e.g. water, soils, or minerals agricultural products, are crucial in triggering economic growth; production processes, however, require several sources of energy, in particular, fossil fuel (Aslam et al., 2021; Salvati & Carlucci, 2011; Wilson et al., 2017). The massive adoption of fossil fuel combustion has sharply increased atmospheric carbon dioxide concentrations (Incerti et al., 2007; Rahman et al., 2022; Salvati et al., 2012).

Carbon dioxide (CO_2) is a natural gas and a by-product of fossil fuel combustion and biomass combustion (Kazim, 2007; Li & Wang, 2023; Vicedo-Cabrera et al., 2018), spreading also from changes in land-use and other industrial processes (Meinshausen et al., 2009; Norby & Luo, 2004; Salvati & Zitti, 2009). The Kyoto Protocol, issued in 2005, recommended a global reduction of CO_2 emissions. The Treaty requires a reduction in emissions of pollutants in the time period between 2008 and 2012, by no less than 5% compared with those recorded in 1990, considered as the base year (Francaviglia et al., 2019; Salvati & Zitti, 2008; Tomao et al., 2017).

While being partly ineffective, this treaty stimulated several agreements on climate change. As a result of an increasing commitment to cut emissions at the different Conferences of the Parties (COP), the goal of sustainable development with zero emissions is drawing nearer. Nations that have ratified the agreement nearly invariably fail to meet their targets. This fact does not imply that emissions have not been decreased, but rather that policymakers have not chosen policies completely coherent with their objectives (Iyer et al., 2015; Mikulčić et al., 2013; Moser, 2010). Since 1997, CO_2 emissions have continued to increase in many countries, despite the Kyoto Protocol emission limits (Aichele et al., 2013; Paraschiv & Paraschiv, 2020; Zickfeld et al., 2009). The most recent climate agreement is the one signed in Paris in 2015 (Huang et al., 2017). It aims at containing the average global surface warming to less than 2 °C compared with pre-industrial levels (Rosković & Bjegović, 2005), through planning, implementation and financing (Lesnikowski et al., 2017).

In general, the expectation of climate change raises many ethical questions (e.g. Seifollahi-Aghmiuni et al., 2022). On the one hand, the envi-

C. Ciaschini and L. Salvati, *Sustainable Development*, Sustainable Development Goals Series,
https://doi.org/10.1007/978-3-032-12559-0_6

ronmental issue of pollution seems obvious and uncontroversial, i.e. pollution degrades the environment (Lau et al., 2012). A closer look, however, reveals several possible interacting factors (Yu et al., 2008). Specifically, at least three alternative visions seem to emerge (McCulloch et al., 2012): (i) the view of 'damage', where pollution devalues the environment, both intrinsically and externally; (ii) the vision of 'transgression', where pollution does not respect and/or violates the rights of others; and (iii) the perspective of 'vice', i.e. pollution is a vice and not the kind of thing that a person of virtuous moral character would do (Zickfeld et al., 2009).

Literature Review

The issue of climate change has been originally spread among scholars since the emergence of a more concrete awareness on greenhouse effect (e.g. Mancino et al., 2016; Roelfsema et al., 2020; Sallustio et al., 2018). Climate change has rapidly revealed a crucial environmental issue with important economic implications deserving a coordinate policy response (Hickman & Banister, 2007; Sterman & Sweeney, 2002; Szabó et al., 2006). The increasing awareness on greenhouse emissions and its consequences led to relevant increases in investments in research and development promoting more sustainable technologies at lower carbon emissions (Holding et al., 2015; Shakun et al., 2012; Singh et al., 2008). These kinds of investments generate several market opportunities, stimulating sectors dealing with renewable energies, energy efficiency and sustainable mobility, and thus creating new job positions and favouring the green economic growth (Egidi et al., 2020; Loarie et al., 2009; Mensah et al., 2018; Zevenhoven & Beyene, 2011). Conversely, climate change also implies significant costs (Hasanbeigi et al., 2012; Hughes et al., 2003; Riebesell et al., 2007). Namely, extreme climate events as floods, drought and storms, cause relevant damages to infrastructures, agriculture fields and productive activities, with direct consequences on GDP and employment (Aubinet et al., 2000; Ho et al., 2011; Su et al., 2021).

Moreover, policies of emissions' reduction, as carbon tax and cap-and-trade systems, could represent both an opportunity and a challenge for firms, forcing them to conform to new standards, e.g. increasing the level of investments in sustainable technologies (Ferrara et al., 2017; Grunewald & Martinez-Zarzoso, 2016; Lamonica et al., 2020). The balance between such costs and benefits leads the design of economic and environmental policies (Alhindawi et al., 2020; Ali et al., 2011; Clémençon, 2016). A transition towards a greener economy requires initial investments but can also bring long-term benefits, including a reduction of the economic impact of extreme climate events and a higher resilience of the global economic system (Anderson et al., 2016; Kessel, 2000; Yamasaki, 2003). Studying connections between climate change and economy highlights the need of integrated strategies that consider both challenges and opportunities of a sustainable future (Keeling, 1995; Liaquat et al., 2010; Rugani et al., 2014).

Environmental Impacts and Policy Implications

Within the present global framework, the impact of economic and environmental policies on greenhouse emissions represents a topic of crucial relevance characterizing unprecedented socio-ecological challenges (Al-Ghussain, 2019; Kurz & Apps, 2006; Viguier et al., 2003) and requiring an increasingly integrated international commitment to contain the global warming (Benhelal et al., 2013). The existing interdependent relationship between public policies, economic dynamics and CO_2 emissions specifically requires a deep analysis to design effective strategic interventions and lead proper economic behaviours contributing to the reduction of greenhouse gas emissions (Houghton, 2005; Joanes & Gill, 1998; Montzka et al., 2011). At first, environmental policies play a relevant role in driving the transition towards a more sustainable economy (Keeling et al., 2011; Worrell et al., 2001; Xu et al., 2013). Norms and regulations, as the Emission Trading System (ETS), carbon taxes and the standard of energy efficiency, should encourage firms and citizens to reduce their own emissions (Riebesell et al., 2007; Thisted & Thisted, 2020; Wei et al., 2012).

Specifically, the introduction of a price on carbon makes economically discourages the exploitation of fossil sources, rather stimulating the adoption of renewable energies and cleaner technologies (Anwar et al., 2018; Fang et al., 2011; Zhou, 2023). In addition, incentives, as subsidies for renewable energies or financing for research and development on low emission technologies, would favour the emergence and spread of sustainable practices (Oberthür & Groen, 2018; Rosenstiel et al., 2003; Thornton et al., 2014). Nevertheless, the effectiveness of these policies depends on their implementation, and the political acknowledgement of long-term commitment (Day et al., 2008). The present ineffective and weak regulations along with the insufficient incentives could prevent a significant reduction of emissions (Geden, 2016). The dynamics of economic growth, production and consumption directly affect the levels of CO_2 emissions (Falkner, 2016; Nordebo et al., 2020; Weart, 2007).

Expanding economies tend to increase its demand of energy, often deriving from fossil sources, thus contributing to increase emissions (Hoegh-Guldberg & Bruno, 2010). Nevertheless, a more efficient economy, with advanced technologies and sustainable practices, can reduce the emissions even in presence of economic growth (Bashir et al., 2022). Globalization and international commercial flows play a relevant role, since delocalised productions can shift the emissions from a country to another, thus complicating the overall evaluation of climate policies (Khasnis & Nettleman, 2005; Ferrara et al., 2016). In addition, the consumption choices of people and firms are affected by energy prices, wages, their availability of sustainable technologies and environmental awareness (Florides & Christodoulides, 2009). The economic choices and policies reciprocally affect each other. In this sense, effective policies can shape proper economic behaviours, while the market dynamics can require more effective normative interventions (Bodansky, 2016). At the same time, energy transitions from fossil sources to renewable resources, require both economic incentives and a favourable regulatory framework.

In general, the impact of economic policies on CO_2 emissions is deep and trivial (Caldeira et al., 2003). Properly designed and effectively implemented policies can lead the economic behaviours towards more sustainable practices thus significantly reducing emissions (Gao et al., 2012). At the same time, economic dynamics, affected by growth, innovation and consumption choices, can facilitate or prevent this transition (Chapman, 2007). An integrated approach is thus crucial to effectively assess climate change (Friedlingstein et al., 2014), combining effective policy interventions with economic strategies oriented to sustainability (Cherubini et al., 2011).

Empirical Findings

The work, focussing on per capita CO_2 emissions over 58 years, relies on a sample of 266 countries retrieved from World Bank database. Nevertheless, countries characterized by a highly discontinuous historical series have been neglected. In addition, multivariate analysis has been performed on the first differences of the available time series starts from 1961. All these data elaborations lead to a final database of 151 rows (countries) and 57 columns (years). Throughout the years, CO_2 emissions experienced several changes. CO_2 emissions in world countries reflect two different spatial contexts existing in 1960: (i) values above 5 metric tons per capita in industrialized countries and emerging economies, (ii) values below 5 metric tons per capita in economically disadvantaged countries such as in Africa and Latin America. For instance, emissions in the United States exceeded 16 metric tons per capita due to a highly flourishing economic situation and its dominant position within the global economy. Since the early 1990s, many countries, including Germany, began to collect data on CO_2 emissions, possibly highlighting a more sustainable development path in the last year of analysis. In 2018, global CO_2 emissions resulted to be moderately contained, below 16 metric tons per capita on average.

Descriptive Statistics

Table 6.1 reports the individual value of descriptive statistics for each decade under scrutiny. In 1960, on average, global CO_2 emissions seem to be lower than in other decades. The highest average values were observed in the 1970s and the 1980s, while constant emissions have been shown in the years of the twenty-first century, with a slight reversal between 2010 and 2018. Descriptive statistics show a decreasing asymmetry over time, at least until the 1990s. From this decade onwards, skewness rests around the range 2.3–2.6. In 1960, a very high kurtosis index (32) may reveals outlier countries with very low emissions. Kurtosis, like skewness, decreased progressively until 1990, ranging between 6.3 and 8.6. In addition, years 1990 and 2018 show the same value of kurtosis, equal to 7.4. Throughout the period, CO_2 emissions did not perform a constant trend. This path was strongly affected by population dynamics and economic development. Economic growth characterized the initial period, corresponding to the first 10–20 years after World War II. A slight variation in the average CO_2 in the first three decades, from 2.0 to 2.2 metric tons per capita, was observed. A sudden increase to 3.0 metric tons per capita reflected in higher fluctuations across countries since 1963. In 1969, the global average CO_2 emissions rose to about 4.5 metric tons per capita. Only four years later, this value approached 5.2 metric tons.

In the following years, emissions started a decreasing trend, laying below 4.5 metric tons per capita until 1989. From 1990 onwards, global average value of carbon dioxide emissions began a new (slow but steady) increase, which ended in 2006. In these 16 years, the average value ranged between 3.7 and 4.5 metric tons per capita, with constant and lower standard errors compared with those exhibited between 1963 and 1980. An opposite trend has been observed from 2007, possibly reflecting an increased awareness of environmentally sustainable behaviours within populations. In these years, more decisive approaches towards the reduction in CO_2 emissions have been acknowledged. Nevertheless, some peculiarities emerged in 2009 and 2012, where a slight increase may relate to global economic recession. In this perspective, a particularly high level of variability was observed in the 1970s and the 1980s, in line with pronounced inequalities within the 151 countries of the sample. In these decades, indeed, several countries reached average CO_2 emissions close to 70 metric tons per capita.

Conversely, lower degrees of inequalities have been highlighted between more polluting and less polluting countries in more recent decades. Peculiarities emerge in the case of Luxembourg, with significantly lower emissions in 2018 compared with 1960. On the contrary, Middle-East countries, as Qatar, the Arab Emirates and Saudi Arabia, seem to have higher emissions in 2018 than in 1960. This outcome should be reasonably justified with the fact that the economies of such countries—rich in oil fields—sharply developed in recent decades. The activity of developed economies in turn translated into higher values of

Table 6.1 Descriptive statistics representing the distribution of CO_2 emissions (metric tons per capita) across world countries

Metric	1960	1970	1980	1990	2000	2010	2018
Mean	2	4.6	4.8	3.8	4.2	4.4	4.1
Minimum	0.01	0.02	0.04	0	0.02	0.03	0.03
Maximum	36.6	69.1	58.5	30.3	36.7	33.5	32.4
Norm. range	18.1	14.7	12.0	7.9	8.7	7.6	7.9
Median	0.4	0.9	1.32	1.24	1.82	2.31	2.2
25 percentile	0.1	0.3	0.4	0.3	0.5	0.6	0.6
75 percentile	1.7	5.5	6.6	6.1	6.1	6.2	5.8
Skewness	5.1	4.5	3.3	2.5	2.6	2.3	2.4
Kurtosis	32.2	22.8	14.1	7.5	8.6	6.4	7.5

Source Authors' elaboration of World Bank indicators

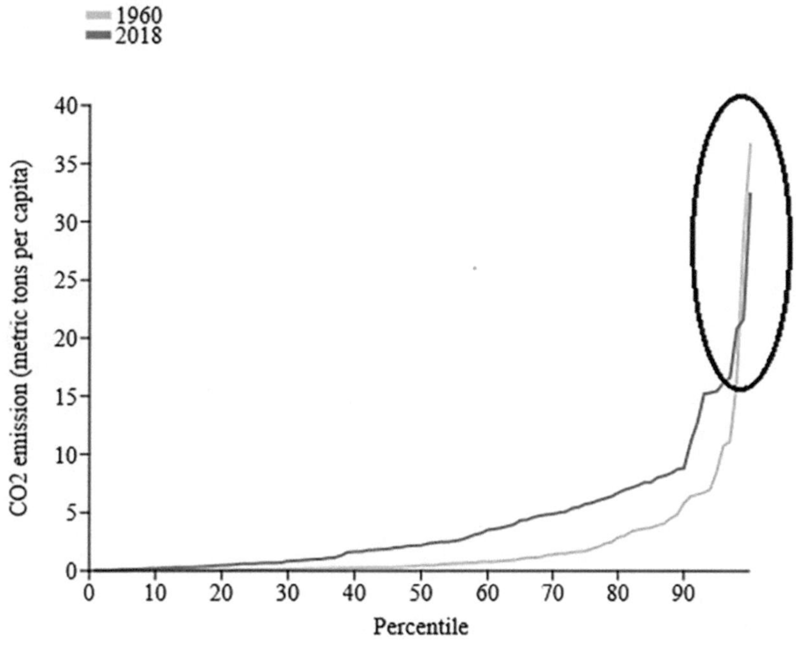

Fig. 6.1 A graphical summary of the statistical distribution of CO_2 emission (metric tons per capita) for 1960 and 2018 based on a percentile plot. *Source* Authors' elaboration of World Bank indicators

per capita CO_2 emissions. Among these countries, Qatar and the Arab Emirates (showing average values of CO_2 emissions of, respectively, 3.7 and 0.1 metric tons per capita in 1960), reached 32.4 and 20.8 metric tons per capita, respectively, in 2018. A more homogeneous path in 2018 emissions than in 1960, is highlighted in Fig. 6.1, comparing the percentiles of CO_2 emissions for the two years. This trend, highlighting a higher level of per capita emissions in 2018, persists until the 98–99% percentile, where about 2% of countries strongly affect emissions to exceed those of 2018.

Correlation Analysis

Figure 6.2 shows trends over time in both Pearson and Spearman coefficients, respectively, reflecting parametric (linear) and non-parametric (non-linear) linkages among data within the time series. Both coefficients exhibit a fluctuating trend throughout the whole time period. While Spearman coefficient seems to be more stable, ranging between −0.2 and 0.4, Pearson values fluctuate in the range between −0.9 and 0.9. In general, the two coefficients exhibit very different trends; nevertheless, in some years, and especially from 2008, they follow a similar pattern even diverging in intensity.

Multidimensional Analysis

A Principal Component Analysis (PCA) run on the same data extracted the first two dimensions accounting, respectively, for 85% and 5% of the total variance, indicating that most of the spatial variation in CO_2 emissions was captured by single dominant factors. Components from 3 onwards only explain a negligible part of the total variance. The PCA biplot in Fig. 6.3 illustrates the relationship between countries (and, indirectly, world regions) based on their CO_2 emissions. The largest variation is captured along the first dimension, while the second axis adds secondary characteristics like energy efficiency and renewable policies.

Points clustered near the origin likely represent countries with average or balanced emissions profiles, while outliers like Qatar (QAT) and Kuwait (KWT) represent unique, high-emission cases. This plot underscores the heterogeneity in CO_2 emissions across regions, driven by factors

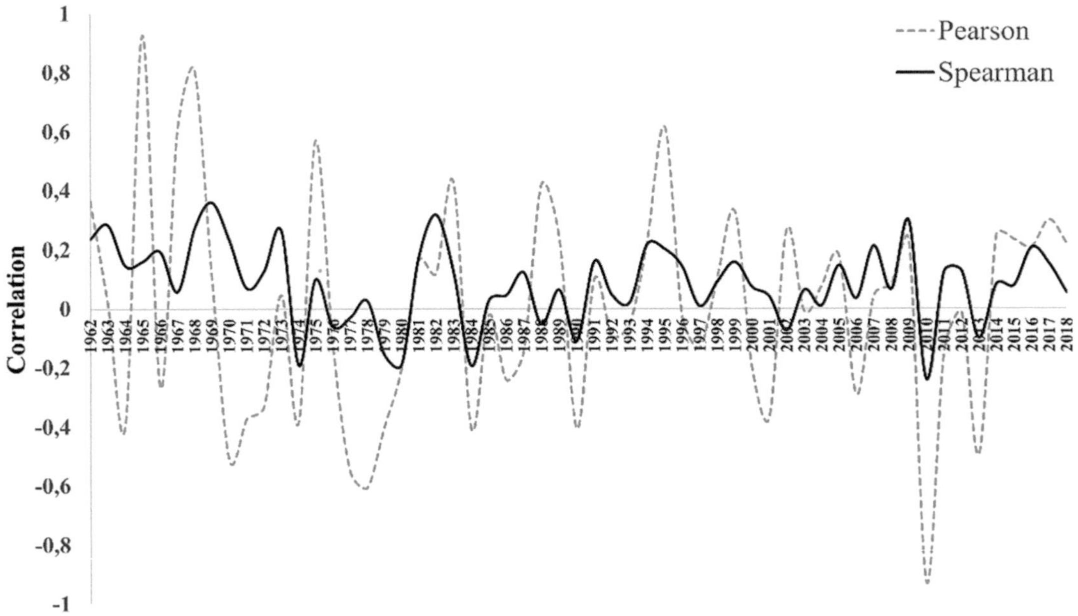

Fig. 6.2 Trends over time in parametric (Pearson) and non-parametric (Spearman) correlation coefficients within the (−1) lagged time series of metric tons per capita CO_2. *Source* Authors' elaboration of World Bank indicators

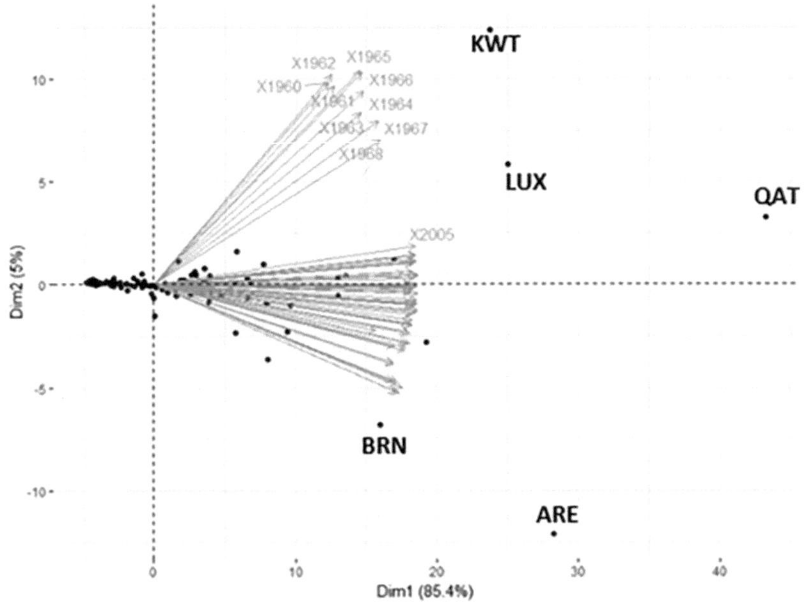

Fig. 6.3 Results of a principal component analysis illustrated in a biplot reflecting the distribution of variables' (years) loadings and rows (countries) scores along the first two axes. *Source* Authors' elaboration of World Bank indicators

such as industrial activity, energy reliance, and policy frameworks. Qatar's position far on the positive side of Component 1 indicates extremely high total CO_2 emissions. Qatar, indeed, is one of the major fossil fuel producers and exporters, characterized by a resource-driven economy. The slightly negative Component 2 value might reflect limited adoption of energy-efficient or renewable practices, contributing to emissions' inefficiencies. Qatar (QAT) stands out for its extraordinarily high emissions, likely driven by its heavy reliance on fossil fuel production and energy-intensive industries. For instance, the positive association of the United Arab Emirates (ARE) with Component 1 indicates significant emissions, though not as extreme as Qatar's. The negative Component 2 value in turn results from a high per capita energy consumption due to outdated energy practices, slowing transition to greener technologies. These outcomes overall depict a rapidly industrializing country heavily reliant on oil exports.

Kuwait's moderately positive Component 1 suggests significant emissions, though lower than Qatar (QAT) or the United Arab Emirates (ARE). A high positive Component 2 value indicates distinct emission-related factors, such as per capita emissions, demographic characteristics, or targeted policies (e.g. energy subsidies or renewables). From this point of view, Kuwait (KWT) has a unique emissions' profile, possibly driven by high per capita emissions coupled with efforts or circumstances (e.g. smaller population or industrial base) that differentiate its emissions' level from others. The slightly negative Component 1 score suggests a relatively low level of total emissions, likely due to the country's smaller economy or less energy-intensive industries. The strongly negative Component 2 value reflects inefficiencies or reliance on traditional energy sources, potentially pointing to low adoption of renewable energy or energy-saving measures. Brunei Darussalam (BRN) performs a contained overall emissions footprint while showing inefficient energy practices. On the contrary, Luxembourg's moderately positive Component 1 indicates high emissions typical of industrialized nations with energy-demanding sectors. The neutral Component 2 score suggests no strong distinguishing secondary factors, implying balanced energy practices or policies. Luxembourg (LUX) represents an average industrialized region with significant emissions, moderate inefficiencies and some recent technological advancements.

Discussion

The increased levels of carbon dioxide emissions represent a crucial challenge to global sustainable development (Godil et al., 2021), negatively affecting climate with phenomena such as tornadoes, snowfalls or the melting of glaciers due to higher temperatures, causing the rise of the seas (Glaeser & Kahn, 2010). Moreover, it also affects plants and animal lives, leading to biodiversity loss (Aldy, 2006). This study, focussing on per capita CO_2 emissions over 58 years, has been complemented with a parallel scrutiny of population growth in the different countries (Wagner et al., 2016).

This approach is particularly helpful for policy implementation because the decrease in per capita emissions, at least in some countries, does not necessarily imply a decrease in the absolute value of emissions (Yang et al., 2023). This trend, indeed, could be also connected with a relevant increase in population overcoming that of emissions (Polasky et al., 2019). The analysed indicator succeeds in the purpose of monitoring trends and highlighting peculiarities connected to global warming (Chen et al., 2019; Hansen et al., 2000; Su et al., 2021). Nevertheless, climate improvement should now benefit from new environmentally sustainable technological innovations (Bruckner et al., 2022; Rehman et al., 2021; Ulucak, 2021). With reference to the healthcare sector (e.g. Song & Tong, 2022), a more sustainable exploitation of natural resources crucially affects the quality of life and have significative effect in the decrease of public healthcare expenditures (Apergis et al., 2018; Khan, 2019; McAlister et al., 2022).

Concluding Remarks

Within a context of sustainable development, persisting high values of carbon dioxide emissions hinder the attainment of most of the Goals stated in the United Nations (UN) agenda, having negative repercussions on specific aspects such as poverty, employment, food provision, and energy resources. Effective policy actions should therefore reduce emissions, thus containing the negative consequences of climate change.

References

Aichele, R., & Felbermayr, G. (2013). The effect of the Kyoto Protocol on carbon emissions. *Journal of Policy Analysis and Management, 32*(4), 731–757.

Aldy, J. E. (2006). Per capita carbon dioxide emissions: Convergence or divergence? *Environmental and Resource Economics, 33*(4), 533–555.

Al-Ghussain, L. (2019). Global warming: Review on driving forces and mitigation. *Environmental Progress and Sustainable Energy, 38*(1), 13–21.

Alhindawi, R., Abu Nahleh, Y., Kumar, A., & Shiwakoti, N. (2020). Projection of greenhouse gas emissions for the road transport sector based on multivariate regression and the double exponential smoothing model. *Sustainability, 12*(21), 9152.

Ali, M. B., Saidur, R., & Hossain, M. S. (2011). A review on emission analysis in cement industries. *Renewable and Sustainable Energy Reviews, 15*(5), 2252–2261.

Anderson, T. R., Hawkins, E., & Jones, P. D. (2016). CO2, the greenhouse effect and global warming: From the pioneering work of Arrhenius and Callendar to today's earth system models. *Endeavour, 40*(3), 178–187.

Anwar, M. N., Fayyaz, A., Sohail, N. F., Khokhar, M. F., Baqar, M., Khan, W. D., & Nizami, A. S. (2018). CO_2 capture and storage: A way forward for sustainable environment. *Journal of Environmental Management, 226*, 131–144.

Apergis, N., Gupta, R., Lau, C. K. M., & Mukherjee, Z. (2018). US state-level carbon dioxide emissions: Does it affect health care expenditure? *Renewable and Sustainable Energy Reviews, 91*, 521–530.

Aslam, B., Hu, J., Shahab, S., Ahmad, A., Saleem, M., Shah, S. S. A., & Hassan, M. (2021). The nexus of industrialization, GDP per capita and CO_2 emission in China. *Environmental Technology and Innovation, 23*, Article 101674.

Aubinet, M., Grelle, A., Ibrom, A., Rannik, Ü., Moncrieff, J., Foken, T., & Vesala, T. (2000). Estimates of the annual net carbon and water exchange of forests: The EUROFLUX methodology. *Advances in Ecological Research, 30*, 113–175.

Barbier, E. B., & Burgess, J. C. (2022). Climate and development: The role of the sustainable development goals. *Climate and Development*, 67–90.

Bashir, M. F., Ma, B., Bashir, M. A., Radulescu, M., & Shahzad, U. (2022). Investigating the role of environmental taxes and regulations for renewable energy consumption: Evidence from developed economies. *Economic Research-Ekonomska Istraživanja, 35*(1), 1262–1284.

Benhelal, E., Zahedi, G., Shamsaei, E., & Bahadori, A. (2013). Global strategies and potentials to curb CO_2 emissions in cement industry. *Journal of Cleaner Production, 51*, 142–161.

Bodansky, D. (2016). The Paris climate change agreement: A new hope? *American Journal of International Law, 110*(2), 288–319.

Bruckner, B., Hubacek, K., Shan, Y., Zhong, H., & Feng, K. (2022). Impacts of poverty alleviation on national and global carbon emissions. *Nature Sustainability, 5*(4), 311–320.

Caldeira, K., Jain, A. K., & Hoffert, M. I. (2003). Climate sensitivity uncertainty and the need for energy without CO_2 emission. *Science, 299*(5615), 2052–2054.

Chapman, L. (2007). Transport and climate change: A review. *Journal of Transport Geography, 15*(5), 354–367.

Chen, L., Zhuo, Y., Xu, Z., Xu, X., & Gao, X. (2019). Is carbon dioxide (CO_2) emission an important factor affecting healthcare expenditure? Evidence from China, 2005–2016. *International Journal of Environmental Research and Public Health, 16*(20), 3995.

Cherubini, F., Peters, G. P., Berntsen, T., Strømman, A. H., & Hertwich, E. (2011). CO_2 emissions from biomass combustion for bioenergy: Atmospheric decay and contribution to global warming. *Gcb Bioenergy, 3*(5), 413–426.

Clémençon, R. (2016). The two sides of the Paris climate agreement: Dismal failure or historic breakthrough? *The Journal of Environment & Development, 25*(1), 3–24.

Day, J. W., Christian, R. R., Boesch, D. M., Yáñez-Arancibia, A., Morris, J., Twilley, R. R., & Schaffner, L. (2008). Consequences of climate change on the eco geomorphology of coastal wetlands. *Estuaries and Coasts, 31*(3), 477–491.

Egidi, G., Salvati, L., & Vinci, S. (2020). The long way to tipperary: City size and worldwide urban population trends, 1950–2030. *Sustainable Cities and Society, 60*, Article 102148.

Falkner, R. (2016). The Paris agreement and the new logic of international climate politics. *International Affairs, 92*(5), 1107–1125.

Fang, J., Zhu, J., Wang, S., Yue, C., & Shen, H. (2011). Global warming, human-induced carbon emissions, and their uncertainties. *Science China Earth Sciences, 54*, 1458–1468.

Ferrara, A., Kelly, C., Wilson, G. A., Nolè, A., Mancino, G., Bajocco, S., & Salvati, L. (2016). Shaping the role of 'fast' and 'slow' drivers of change in forest-shrubland socio-ecological systems. *Journal of Environmental Management, 169*, 155–166.

Ferrara, C., Carlucci, M., Grigoriadis, E., Corona, P., & Salvati, L. (2017). A comprehensive insight into the geography of forest cover in Italy: Exploring the importance of socioeconomic local contexts. *Forest Policy and Economics, 75*, 12–22.

Florides, G. A., & Christodoulides, P. (2009). Global warming and carbon dioxide through sciences. *Environment International, 35*(2), 390–401.

Francaviglia, R., Di Bene, C., Farina, R., Salvati, L., & Vicente-Vicente, J. L. (2019). Assessing "4 per 1000" soil organic carbon storage rates under Mediterranean climate: A comprehensive data analysis. *Mitigation and Adaptation Strategies for Global Change, 24*, 795–818.

Friedlingstein, P., Andrew, R. M., Rogelj, J., Peters, G. P., Canadell, J. G., Knutti, R., & Le Quéré, C. (2014).

Persistent growth of CO_2 emissions and implications for reaching climate targets. *Nature Geoscience, 7*(10), 709–715.

Gao, K., Xu, J., Gao, G., Li, Y., Hutchins, D. A., Huang, B., & Riebesell, U. (2012). Rising CO_2 and increased light exposure synergistically reduce marine primary productivity. *Nature Climate Change, 2*(7), 519–523.

Geden, O. (2016). The Paris agreement and the inherent inconsistency of climate policymaking. *Wiley Interdisciplinary Reviews: Climate Change, 7*(6), 790–797.

Glaeser, E. L., & Kahn, M. E. (2010). The greenness of cities: Carbon dioxide emissions and urban development. *Journal of Urban Economics, 67*(3), 404–418.

Godil, D. I., Yu, Z., Sharif, A., Usman, R., & Khan, S. A. R. (2021). Investigate the role of technology innovation and renewable energy in reducing transport sector CO_2 emission in China: A path toward sustainable development. *Sustainable Development, 29*(4), 694–707.

Grunewald, N., & Martinez-Zarzoso, I. (2016). Did the Kyoto protocol fail? An evaluation of the effect of the Kyoto Protocol on CO_2 emissions. *Environment and Development Economics, 21*(1), 1–22.

Hansen, J., Sato, M., Ruedy, R., Lacis, A., & Oinas, V. (2000). Global warming in the twenty-first century: An alternative scenario. *Proceedings of the National Academy of Sciences, 97*(18), 9875–9880.

Hasanbeigi, A., Price, L., & Lin, E. (2012). Emerging energy-efficiency and CO_2 emission-reduction technologies for cement and concrete production: A technical review. *Renewable and Sustainable Energy Reviews, 16*(8), 6220–6238.

Hickman, R., & Banister, D. (2007). Looking over the horizon: Transport and reduced CO_2 emissions in the UK by 2030. *Transport Policy, 14*(5), 377–387.

Ho, S. H., Chen, C. Y., Lee, D. J., & Chang, J. S. (2011). Perspectives on microalgal CO_2-emission mitigation systems—A review. *Biotechnology Advances, 29*(2), 189–198.

Hoegh-Guldberg, O., & Bruno, J. F. (2010). The impact of climate change on the world's marine ecosystems. *Science, 328*(5985), 1523–1528.

Holding, J. M., Duarte, C. M., Sanz-Martín, M., Mesa, E., Arrieta, J. M., Chierici, M., & Agusti, S. (2015). Temperature dependence of CO_2-enhanced primary production in the European Arctic Ocean. *Nature Climate Change, 5*(12), 1079–1082.

Houghton, J. (2005). Global warming. *Reports on Progress in Physics, 68*(6), 1343.

Huang, J., Yu, H., Dai, A., Wei, Y., & Kang, L. (2017). Drylands face potential threat under 2 C global warming target. *Nature Climate Change, 7*(6), 417–422.

Hughes, T. P., Baird, A. H., Bellwood, D. R., Card, M., Connolly, S. R., Folke, C., & Roughgarden, J. (2003). Climate change, human impacts, and the resilience of coral reefs. *Science, 301*(5635), 929–933.

Incerti, G., Feoli, E., Salvati, L., Brunetti, A., & Giovacchini, A. (2007). Analysis of bioclimatic time series and their neural network-based classification to characterise drought risk patterns in South Italy. *International Journal of Biometeorology, 51*, 253–263.

Iyer, G. C., Edmonds, J. A., Fawcett, A. A., Hultman, N. E., Alsalam, J., Asrar, G. R., & McJeon, H. C. (2015). The contribution of Paris to limit global warming to 2 C. *Environmental Research Letters, 10*(12), Article 125002.

Joanes, D. N., & Gill, C. A. (1998). Comparing measures of sample skewness and kurtosis. *Journal of the Royal Statistical Society: Series D (The Statistician), 47*(1), 183–189.

Kazim, A. M. (2007). Assessments of primary energy consumption and its environmental consequences in the United Arab Emirates. *Renewable and Sustainable Energy Reviews, 11*(3), 426–446.

Keeling, C. D., Piper, S. C., Whorf, T. P., & Keeling, R. F. (2011). Evolution of natural and anthropogenic fluxes of atmospheric CO_2 from 1957 to 2003. *Tellus b: Chemical and Physical Meteorology, 63*(1), 1–22.

Keeling, C. D., Whorf, T. P., Wahlen, M., & Van der Plichtt, J. (1995). Interannual extremes in the rate of rise of atmospheric carbon dioxide since 1980. *Nature, 375*(6533), 666–670.

Kessel, D. G. (2000). Global warming—Facts, assessment, countermeasures. *Journal of Petroleum Science and Engineering, 26*(1–4), 157–168.

Khan, S. A. R. (2019). The nexus between carbon emissions, poverty, economic growth, and logistics operations-empirical evidence from southeast Asian countries. *Environmental Science and Pollution Research, 26*(13), 13210–13220.

Khasnis, A. A., & Nettleman, M. D. (2005). Global warming and infectious disease. *Archives of Medical Research, 36*(6), 689–696.

Kurz, W. A., & Apps, M. J. (2006). Developing Canada's national forest carbon monitoring, accounting and reporting system to meet the reporting requirements of the Kyoto Protocol. *Mitigation and Adaptation Strategies for Global Change, 11*(1), 33–43.

Lamonica, G. R., Recchioni, M. C., Chelli, F. M., & Salvati, L. (2020). The efficiency of the cross-entropy method when estimating the technical coefficients of input–output tables. *Spatial Economic Analysis, 15*(1), 62–91.

Lau, L. C., Lee, K. T., & Mohamed, A. R. (2012). Global warming mitigation and renewable energy policy development from the Kyoto Protocol to the Copenhagen Accord—A comment. *Renewable and Sustainable Energy Reviews, 16*(7), 5280–5284.

Lesnikowski, A., Ford, J., Biesbroek, R., Berrang-Ford, L., Maillet, M., Araos, M., & Austin, S. E. (2017). What does the Paris agreement mean for adaptation? *Climate Policy, 17*(7), 825–831.

Li, R., & Wang, Q. (2023). Does renewable energy reduce per capita carbon emissions and per capita ecological footprint? New evidence from 130 countries. *Energy Strategy Reviews, 49*, Article 101121.

Liaquat, A. M., Kalam, M. A., Masjuki, H. H., & Jayed, M. H. (2010). Potential emissions reduction in road transport sector using biofuel in developing countries. *Atmospheric Environment, 44*(32), 3869–3877.

Loarie, S. R., Duffy, P. B., Hamilton, H., Asner, G. P., Field, C. B., & Ackerly, D. D. (2009). The velocity of climate change. *Nature, 462*(7276), 1052–1055.

Maâlej, A., & Cabagnols, A. (2020). CO_2 emissions and growth. *The Journal of Energy and Development, 46*(1/2), 1–24.

Mancino, G., Nolè, A., Salvati, L., & Ferrara, A. (2016). In-between forest expansion and cropland decline: A revised USLE model for soil erosion risk under land-use change in a Mediterranean region. *Ecological Indicators, 71*, 544–550.

McAlister, S., Morton, R. L., & Barratt, A. (2022). Incorporating carbon into health care: Adding carbon emissions to health technology assessments. *The Lancet Planetary Health, 6*(12), e993–e999.

McCulloch, M., Falter, J., Trotter, J., & Montagna, P. (2012). Coral resilience to ocean acidification and global warming through pH up-regulation. *Nature Climate Change, 2*(8), 623–627.

Meinshausen, M., Meinshausen, N., Hare, W., Raper, S. C., Frieler, K., Knutti, R., & Allen, M. R. (2009). Greenhouse-gas emission targets for limiting global warming to 2 C. *Nature, 458*(7242), 1158–1162.

Mensah, C. N., Long, X., Boamah, K. B., Bediako, I. A., Dauda, L., & Salman, M. (2018). The effect of innovation on CO_2 emissions of OCED countries from 1990 to 2014. *Environmental Science and Pollution Research, 25*(29), 29678–29698.

Mikulčić, H., Vujanović, M., Markovska, N., Filkoski, R. V., Ban, M., & Duić, N. (2013). CO_2 emission reduction in the cement industry. *Chemical Engineering Transactions, 35*, 703–708.

Montzka, S. A., Dlugokencky, E. J., & Butler, J. H. (2011). Non-CO_2 greenhouse gases and climate change. *Nature, 476*(7358), 43–50.

Moser, S. C. (2010). Communicating climate change: History, challenges, process and future directions. *Wiley Interdisciplinary Reviews: Climate Change, 1*(1), 31–53.

Norby, R. J., & Luo, Y. (2004). Evaluating ecosystem responses to rising atmospheric CO_2 and global warming in a multi-factor world. *New Phytologist, 162*(2), 281–293.

Nordebo, S., Naeem, M. F., & Tans, P. (2020). Estimating the short-time rate of change in the trend of the Keeling curve. *Scientific Reports, 10*(1), 21222.

Oberthür, S., & Groen, L. (2018). Explaining goal achievement in international negotiations: The EU and the Paris Agreement on climate change. *Journal of European Public Policy, 25*(5), 708–727.

Paraschiv, S., & Paraschiv, L. S. (2020). Trends of carbon dioxide (CO_2) emissions from fossil fuels combustion (coal, gas and oil) in the EU member states from 1960 to 2018. *Energy Reports, 6*, 237–242.

Pecl, G. T., Araújo, M. B., Bell, J. D., Blanchard, J., Bonebrake, T. C., Chen, I. C., & Williams, S. E. (2017). Biodiversity redistribution under climate change: Impacts on ecosystems and human well-being. *Science, 355*(6332), eaai9214.

Polasky, S., Kling, C. L., Levin, S. A., Carpenter, S. R., Daily, G. C., Ehrlich, P. R., & Lubchenco, J. (2019). Role of economics in analyzing the environment and sustainable development. *Proceedings of the National Academy of Sciences, 116*(12), 5233–5238.

Radhi, H. (2009). Evaluating the potential impact of global warming on the UAE residential buildings—A contribution to reduce the CO_2 emissions. *Building and Environment, 44*(12), 2451–2462.

Rahman, M. M., Ahmed, R., Mashud, A. H. M., Malik, A. I., Miah, S., & Abedin, M. Z. (2022). Consumption-based CO_2 emissions on sustainable development goals of SAARC region. *Sustainability, 14*(3), 1467.

Rehman, A., Ma, H., Ozturk, I., Murshed, M., & Dagar, V. (2021). The dynamic impacts of CO_2 emissions from different sources on Pakistan's economic progress: A roadmap to sustainable development. *Environment, Development and Sustainability, 23*(12), 17857–17880.

Reilly, J., Mayer, M., & Harnisch, J. (2002). The Kyoto Protocol and non-CO_2 greenhouse gases and carbon sinks. *Environmental Modeling & Assessment, 7*, 217–229.

Riebesell, U., Schulz, K. G., Bellerby, R. G. J., Botros, M., Fritsche, P., Meyerhöfer, M., & Zöllner, E. (2007). Enhanced biological carbon consumption in a high CO_2 ocean. *Nature, 450*(7169), 545–548.

Roelfsema, M., Van Soest, H. L., Harmsen, M., Van Vuuren, D. P., Bertram, C., den Elzen, M., & Vishwanathan, S. S. (2020). Taking stock of national climate policies to evaluate implementation of the Paris Agreement. *Nature Communications, 11*(1), 1–12.

Rosenstiel, T. N., Potosnak, M. J., Griffin, K. L., Fall, R., & Monson, R. K. (2003). Increased CO_2 uncouples growth from isoprene emission in an agriforest ecosystem. *Nature, 421*(6920), 256–259.

Rosković, R., & Bjegović, D. (2005). Role of mineral additions in reducing CO_2 emission. *Cement and Concrete Research, 35*(5), 974–978.

Rugani, B., Roviani, D., Hild, P., Schmitt, B., & Benetto, E. (2014). Ecological deficit and use of natural capital in Luxembourg from 1995 to 2009. *Science of the Total Environment, 468*, 292–301.

Sallustio, L., Pettenella, D., Merlini, P., Romano, R., Salvati, L., Marchetti, M., & Corona, P. (2018). Assessing the economic marginality of agricultural lands in Italy to support land use planning. *Land Use Policy, 76*, 526–534.

Salvati, L., & Carlucci, M. (2011). The economic and environmental performances of rural districts in Italy: Are competitiveness and sustainability compatible targets? *Ecological Economics, 70*(12), 2446–2453.

Salvati, L., & Carlucci, M. (2014). A composite index of sustainable development at the local scale: Italy as a case study. *Ecological Indicators, 43*, 162–171.

Salvati, L., Perini, L., Sabbi, A., & Bajocco, S. (2012). Climate aridity and land use changes: A regional-scale analysis. *Geographical Research, 50*(2), 193–203.

Salvati, L., & Zitti, M. (2008). Regional convergence of environmental variables: Empirical evidences from land degradation. *Ecological Economics, 68*(1–2), 162–168.

Salvati, L., & Zitti, M. (2009). Substitutability and weighting of ecological and economic indicators: Exploring the importance of various components of a synthetic index. *Ecological Economics, 68*(4), 1093–1099.

Seifollahi-Aghmiuni, S., Kalantari, Z., Egidi, G., Gaburova, L., & Salvati, L. (2022). Urbanisation-driven land degradation and socioeconomic challenges in peri-urban areas: Insights from Southern Europe. *Ambio, 51*(6), 1446–1458.

Shakun, J. D., Clark, P. U., He, F., Marcott, S. A., Mix, A. C., Liu, Z., & Bard, E. (2012). Global warming preceded by increasing carbon dioxide concentrations during the last deglaciation. *Nature, 484*(7392), 49–54.

Singh, A., Gangopadhyay, S., Nanda, P. K., Bhattacharya, S., Sharma, C., & Bhan, C. (2008). Trends of greenhouse gas emissions from the road transport sector in India. *Science of the Total Environment, 390*(1), 124–131.

Song, Z., & Tong, P. (2022). The impact of environmental regulation on human sustainable development: Evidence from China. *Sustainability, 14*(19), 11992.

Sterman, J. D., & Sweeney, L. B. (2002). Cloudy skies: Assessing public understanding of global warming. *System Dynamics Review: THe Journal of the System Dynamics Society, 18*(2), 207–240.

Su, C. W., Xie, Y., Shahab, S., Faisal, C. M. N., Hafeez, M., & Qamri, G. M. (2021). Towards achieving sustainable development: Role of technology innovation, technology adoption and CO_2 emission for BRICS. *International Journal of Environmental Research and Public Health, 18*(1), 277.

Szabó, L., Hidalgo, I., Ciscar, J. C., & Soria, A. (2006). CO_2 emission trading within the European Union and Annex B countries: The cement industry case. *Energy Policy, 34*(1), 72–87.

Thisted, E. V., & Thisted, R. V. (2020). The diffusion of carbon taxes and emission trading schemes: The emerging norm of carbon pricing. *Environmental Politics, 29*(5), 804–824.

Thornton, P. K., Ericksen, P. J., Herrero, M., & Challinor, A. J. (2014). Climate variability and vulnerability to climate change: A review. *Global Change Biology, 20*(11), 3313–3328.

Tomao, A., Quatrini, V., Corona, P., Ferrara, A., Lafortezza, R., & Salvati, L. (2017). Resilient landscapes in Mediterranean urban areas: Understanding factors influencing forest trends. *Environmental Research, 156*, 1–9.

Ulucak, R. (2021). Renewable energy, technological innovation and the environment: A novel dynamic auto-regressive distributive lag simulation. *Renewable and Sustainable Energy Reviews, 150*, Article 111433.

Vardopoulos, I., Tsilika, E., Sarantakou, E., Zorpas, A. A., Salvati, L., & Tsartas, P. (2021). An integrated SWOT-PESTLE-AHP model assessing sustainability in adaptive reuse projects. *Applied Sciences, 11*(15), 7134.

Vicedo-Cabrera, A. M., Guo, Y., Sera, F., Huber, V., Schleussner, C. F., Mitchell, D., & Gasparrini, A. (2018). Temperature-related mortality impacts under and beyond Paris Agreement climate change scenarios. *Climatic Change, 150*(3), 391–402.

Viguier, L. L., Babiker, M. H., & Reilly, J. M. (2003). The costs of the Kyoto Protocol in the European Union. *Energy Policy, 31*(5), 459–481.

Wagner, L., Ross, I., Foster, J., & Hankamer, B. (2016). Trading off global fuel supply, CO_2 emissions and sustainable development. *PLoS ONE, 11*(3), Article e0149406.

Weart, S. R. (2007). Money for Keeling: Monitoring CO_2 levels. *Historical Studies in the Physical and Biological Sciences, 37*(2), 435–452.

Wei, T., Yang, S., Moore, J. C., Shi, P., Cui, X., Duan, Q., & Dong, W. (2012). Developed and developing world responsibilities for historical climate change and CO_2 mitigation. *Proceedings of the National Academy of Sciences, 109*(32), 12911–12915.

Wilson, G. A., Kelly, C. L., Briassoulis, H., Ferrara, A., Quaranta, G., Salvia, R., Detsis, V., Curfs, M., Cerda, A., El-Aich, A., Liu, H., Kosmas, C., Alados, C. L., Imeson, A., Landgrebe-Trinkunaite, R., Salvati, L., Naumann, S., Danwen, H., Iosifides, T., … Zhang, P. (2017). Social memory and the resilience of communities affected by land degradation. *Land Degradation and Development, 28*(2), 383–400.

Worrell, E., Price, L., Martin, N., Hendriks, C., & Meida, L. O. (2001). Carbon dioxide emissions from the global cement industry. *Annual Review of Energy and the Environment, 26*(1), 303–329.

Xu, Z., Shimizu, H., Yagasaki, Y., Ito, S., Zheng, Y., & Zhou, G. (2013). Interactive effects of elevated CO_2, drought, and warming on plants. *Journal of Plant Growth Regulation, 32*(4), 692–707.

Yamasaki, A. (2003). An overview of CO_2 mitigation options for global warming—Emphasizing CO_2 sequestration options. *Journal of Chemical Engineering of Japan, 36*(4), 361–375.

Yang, W., Feng, L., Wang, Z., & Fan, X. (2023). Carbon emissions and national sustainable development goals coupling coordination degree study from a global perspective: Characteristics, heterogeneity, and spatial effects. *Sustainability, 15*(11), 9070.

Yu, K. M. K., Curcic, I., Gabriel, J., & Tsang, S. C. E. (2008). Recent advances in CO_2 capture and utilization. *ChemSusChem: Chemistry & Sustainability Energy & Materials, 1*(11), 893–899.

Zevenhoven, R., & Beyene, A. (2011). The relative contribution of waste heat from power plants to global warming. *Energy, 36*(6), 3754–3762.

Zhou, Y. (2023). Worldwide carbon neutrality transition? Energy efficiency, renewable, carbon trading and advanced energy policies. *Energy Reviews, 2*(2), Article 100026.

Zickfeld, K., Eby, M., Matthews, H. D., & Weaver, A. J. (2009). Setting cumulative emissions targets to reduce the risk of dangerous climate change. *Proceedings of the National Academy of Sciences, 106*(38), 16129–16134.

Agricultural Land

7

Introduction

The notion of 'agricultural land', especially within the broader concept of 'land use', basically reflects the share of arable land, permanent crops and grazing in total landscape (Gebremedhin et al., 2018). Abandoned land is usually excluded from this general definition because of the inherent change in cultivation and use, having no (or negligible) economic value (Pongratz et al., 2008). Permanent crops include land cultivated with flowering shrubs, fruit trees, nuts and vines, while excluding woodland and timberland (Eichhorn et al., 2006). At the same time, permanent pastures included into the 'agricultural land' definition are used for five (or more) years for the production of fodder, both natural and cultivated (Strassburg et al., 2014).

The indicator quantifying agricultural land at the country scale originates from the database provided by the United Nations Food and Agriculture Organization (FAO) and derived from official national sources through annual surveys via questionnaires (Bizier et al., 2022; Gennari et al., 2022; Persson et al., 2010). In addition, information can also be retrieved from secondary data sources (official statistics derived from agricultural services and technical offices), ad hoc national surveys/publications and the specific information release of various international organizations (Fritz et al., 2010). Despite the efforts of FAO, standard definitions and homogenous reporting methods can be ensured only partly, thus preventing a full comparability of data and indicators quantifying agricultural land in countries with different climates (Yang et al., 2015).

In the perspective of United Nations (UN) Agenda 2030, rural land use, and especially information on agricultural management, is also crucial to attain multiple Sustainable Development Goals (Moyer & Hedden, 2020). For instance, the amount of agricultural land (and, indirectly, the way any soil is managed and, possibly, protected) directly impacts SDG13 (climate action) and SDG15 (life on land), related to various issues dealing with the preservation of ecosystem and the safeguard of pristine climate conditions (Movilla-Pateiro et al., 2021). It also indirectly involves SDGs more focussed on economic issues (Atukunda et al., 2021), such as SDG12 (sustainable consumption and production) to the achievement of SDG1 (zero poverty) and SDG2 (zero hunger).

Literature Review

Three crucial topics should be addressed when evaluating the relationship between agricultural land stocks and sustainable development (Quintas-Soriano et al., 2016). These include (i) the change of the earth's surface by human action (namely, landscape transformations), (ii) food security and (iii) the effects (and conse-

C. Ciaschini and L. Salvati, *Sustainable Development*, Sustainable Development Goals Series,
https://doi.org/10.1007/978-3-032-12559-0_7

quences) of agricultural land on climate change (Berhane, 2018; Heltberg et al., 2009; Hertel & Lobell, 2014). Throughout the years, agricultural activities have had a massive impact on the ecosystem, especially consuming soil and water for irrigation. Another serious problem relates to the production and use of fertilizers, then released into the atmosphere (Ramankutty et al., 2008). Nevertheless, an expansion of agricultural and pasture lands should meet the food needs of a growing population in the world (Dar & Laxmipathi Gowda, 2013). A sustainable agricultural production system requires a timely data collection focussing on agricultural land (Van Veen, 2005).

At the same time, the protection of ecosystems and natural habitats aimed at mitigating the reduction of biodiversity and the extinction of wild species (Henson & Jaffee, 2006). Within the context of agricultural land management, food security implies the availability of adequate food stocks supporting the constant increase in food consumption, and compensate fluctuations in production and prices (Baldos & Hertel, 2014). In Africa, malnutrition has been steadily increasing in all regions since 2015 (Bello, 2024), partly due to the extreme right-wing political mood favouring oligarchies at the expense of policies to support disadvantaged social classes (Natarajan et al., 2022).

On the one side, such policies should provide food to an increasingly growing population, especially in developing countries (Campbell et al., 2016; Dorward et al., 2004; Zhen & Routray, 2003). On the other side, the quality and safety of the food produced and distributed should be improved substantially through political, economic, environmental and social interventions (Pretty et al., 2003). Sustainable agriculture should provide food security for all (Emina, 2021; Kanter et al., 2016; Langhelle, 1999). A coordinated policy action of different states, governments, and the market, is required to combat food insecurity (Pinstrup-Andersen, 2002; Swaminathan & Bhavani, 2013; Weiss, 1990).

While climate change affects all countries worldwide at different intensities (Aydinalp & Cresser, 2008), agriculture seems to be the most affected sector, especially in tropical regions (Grote, 2014). Crop productivity is likely to change because of these effects (Elliott et al., 2014), which will increasingly generate adverse weather events affecting plant species and diseases threatening livestock (Agarwala et al., 2022; Palmisano et al., 2016; Slater et al., 2007). Areas suitable for the cultivation of the main basic crops could undergo geographical displacement in response to climate change (Kumar et al., 2018; Muluneh, 2021; Wheeler & Von Braun, 2013). Mitigating the negative effects of climate change may support rural development policies aimed at combating poverty and malnutrition, especially in the most disadvantaged world regions completely dependent on agricultural production (Aydinalp et al., 2008; Spiertz, 2009; Amundson et al., 2015).

Agriculture and Food Security

Improving infrastructure, technology, and access to credit for smallholder farmers can transform agriculture into a powerful tool for reducing hunger (Rasul, 2016). Programmes such as the 'Zero Hunger Challenge' and the 'Scaling Up Nutrition Movement' emphasize the importance of sustainable agriculture to achieve global food security (Hamidov et al., 2016; Mugambiwa & Tirivangasi, 2017). The Zero Hunger Challenge initiative implies an integrated collaboration between governments, non-governmental organizations, religious communities, the private sector, and research institutions to eliminate hunger and malnutrition (Bolarinwa et al., 2020; Ingutia & Sumelius, 2022). Integrated agricultural resource management can reduce the need to convert forests into farmland, thus preserving biodiversity (Bindi et al., 2011). Restoring degraded lands could also mitigate climate change isolating large amounts of carbon (Luan et al., 2013). However, better scientific understanding of the processes driving soil and landscape degradation, ultimately leading to early desertification, is needed (Chapman, 2020). In general, ending hunger is an achievable goal within our generation, even requiring an integrated approach combining science, public policy, and international collaboration (Gornall et al., 2010).

Investing in sustainable agriculture, social protection and resilient systems will not only improve food security but also contribute to a more equitable, stable, and prosperous future for all (Chen et al., 2023a, 2023b; Knickel et al., 2018; Mârza et al., 2015). In a context of growing populations, efforts should be addressed to introduce innovations to sustainably boost agricultural production (Piorr, 2003). This involves improving global supply chains, reducing food loss and waste, and ensuring access to nutritious food for those suffering from hunger and malnutrition (Jiang et al., 2013). Achieving these goals requires not only technical solutions and investments but also systemic changes in resource management and global policies (Parry et al., 1999). Along with agriculture, fishery and grazing acquire a rather important, even not crucial, role in facing the issue of global hunger (Nastis et al., 2012). The relative importance of these sectors depends on the geographical, economic, social and environmental context (Pandey et al., 2015).

In this perspective, grazing livestock provides a significant source of food, including milk, meat, and other dairy products, which are rich in protein, fats, and essential micronutrients (Volkov & Fomin, 2018). Livestock grazing systems often constitute the only food production method in arid and semi-arid regions unsuitable for crop farming (Humphreys, 2022; Pham & Smith, 2014; Rhodes, 2014). Moreover, some relevant issues seem to emerge, as soil erosion and loss of biodiversity led by overgrazing and soil compaction (Zhang & Cai, 2011). Land-use conflicts (which marginalize grazing systems especially in developing regions) make these processes even worse (Prabhakar, 2021).

At the same time, fisheries are indispensable for achieving multiple and coherent SDGs, providing essential food, livelihoods, and ecosystem services (Qi et al., 2018). However, ensuring their sustainability is a critical challenge often affected by factors as overfishing, largely impacting long-term sea sustainability (Bununu et al., 2023; Peter & Peng, 2014; Tscharntke et al., 2012). In general, climate change largely impacts fishery and grazing, affecting long-term sea sustainability, warming oceans and rising sea levels, while droughts and extreme weather conditions widely affect grazing systems (Song et al., 2018).

Empirical Findings

The global distribution of agricultural land stock across countries was investigated at two reference dates: 1961 and 2018. In 1961, agriculture accounted for a significant share of the landscape in many world regions, particularly in Africa, Middle East, South Asia and South America. In these regions, the economy was not completely focussed on agriculture, assuring especially local subsistence and the export of raw materials. In Eastern Europe and North America, medium–high levels of agricultural land stock responded to a long tradition of intensive cropping. Conversely, Russia, Canada, Australia and some Asiatic regions, had a lower share of agricultural area, due to unfavourable weather and geographical conditions. In 2018, a completely different geography was observed. Densely populated Asian countries, such as India, Pakistan and Bangladesh, experienced a relevant expansion in agricultural area, along with a contextual growth of demographic pressures that determined an increasing food demand. Conversely, some countries in Central Africa and those under Soviet Union past influence, showed a decrease in the agricultural land stock, possibly connected with early desertification, rural abandonment, social conflicts, or a latent economic transition towards other productive sectors.

In Europe and North America, the situation was rather stable, even with slight variations, likely due to urbanization processes and agricultural policies increasingly oriented towards environmental sustainability. Some areas of Latin America, as Brazil and Argentina, show a slight increase in agricultural area, as a result of local processes of crop intensification, particularly for the production of soy and farming (Ruane & Sonnino, 2011). Taken together, the geography of agricultural land stocks seems to be as strictly connected to socioeconomic, environmental and demographic dynamics. They finally delineate a developing world that requires a constant interaction between productive requirements and those connected to a sustainable management of natural resources.

Descriptive Statistics

The results of descriptive statistics run on the 'agricultural land' indicator highlight peculiar cases deviating from the average behaviour of countries over the years. Table 7.1 provides the most relevant metrics quantifying trends over time in agricultural land stock for all the investigated decades. Empirical findings show that, between 1961 and 2018, the global (average) value of the indicator increased by 0.7%. The highest value of the indicator was observed in correspondence with year 1988. In general, global averages resulted to be rather stable over time. The minimum peak increased slightly from 0.3 to 0.6% over the whole time period. Conversely, the maximum value decreased from 93.4 to 80.8%.

Moreover, the normalized range decreased from 2.5 in 1961 to 2.1 in 2018, indicating a coherent decline in the indicator's spatial variability across the panel of world countries. This may confirm the process of convergence toward similar land-use structures and functions in both advanced and emerging economies. In line with global averages, global medians perform a positive trend over time, even with some minor (or moderate) oscillations. The year 1988 exhibits the highest median value (39), while the lowest median (equal to 36) was observed in 1961, outlining a coherent volatility over time. Throughout the analysed period, the 25th percentile increased by more than 2 points, while the 75th percentile decreased by one point. Skewness in turn decreased significantly (from 0.32 to 0.11) between 1961 and 2018. The lowest value (0.07) was again recorded in 1988, while kurtosis increased progressively, slightly dropping in 2018. A moderate decrease in the world average (−1.1%) was observed between 1988 and 1994, further growing from 1994 onwards and finally decreasing from 2004 to 2007. With global recession, agricultural land stock grew slowly and almost constantly (Fig. 7.1).

Correlation Analysis

Figure 7.2 illustrates the value of correlation coefficients (both Pearson and Spearman) throughout the whole time series. Pearson coefficients were rather positive, between 0.1 and 0.3, with some negative exceptions. Spearman coefficients stabilized around 0.5, decreasing only in the last years. In both 1966 and 1967, Pearson coefficients became negative. The same occurred in 1974 and 1976, 1981, and 1996 (with coefficients approaching −0.4). From 1997 onwards, Pearson coefficients turned out to be positive, even if with some fluctuations. Taken together, the empirical findings of the correlation analysis document a particularly volatile development path,

Table 7.1 Descriptive statistics illustrating the distribution of the per cent share of agricultural land in total landscape across world countries, 1961–2018

Metric	1961	1968	1978	1988	1998	2008	2018
Mean	36.7	37.3	37.5	38.2	37.3	37.1	37.4
Minimum	0.3	0.3	0.4	0.6	0.6	0.5	0.5
Maximum	93.4	90.6	86.6	84.8	85.4	83.8	80.8
Norm. range	2.5	2.4	2.3	2.2	2.3	2.3	2.1
Median	36.0	38.3	38.5	39.1	37.5	38.5	38.3
25th percentile	17.5	18.2	18.4	19.6	18.5	18.7	19.7
75th percentile	55.3	55.3	55.6	56.2	55.2	53.3	54.2
Skewness	0.3	0.2	0.1	0.1	0.1	0.1	0.1
Kurtosis	−0.8	−0.9	−1.0	−1.0	−1.0	−1.0	−1.0

Source Authors' elaboration of World Bank indicators

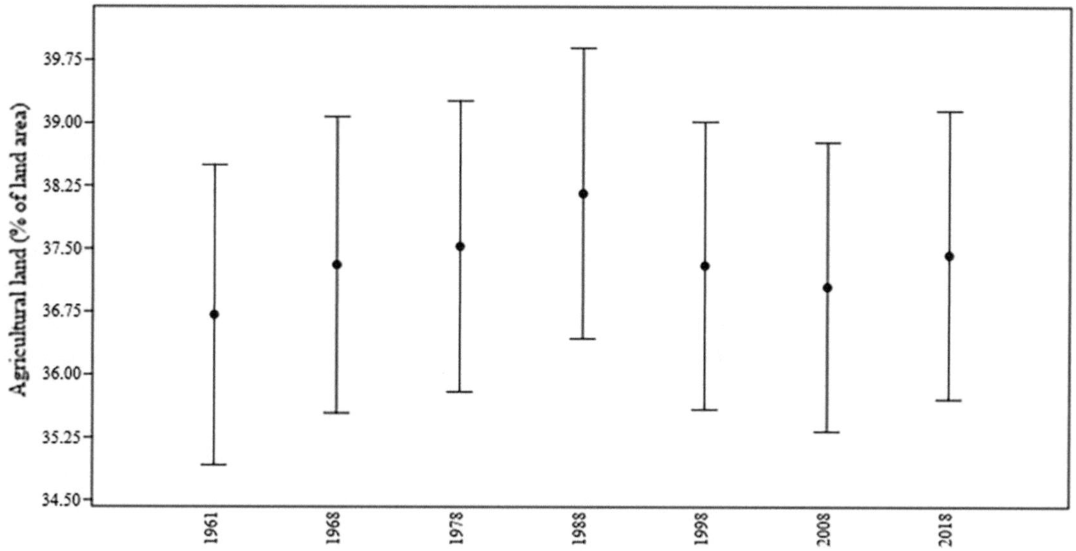

Fig. 7.1 A graphical summary of the spatial distribution of agricultural land (per cent share in total landscape) by decade, based on a mean and whisker plot (standardized data with standard deviation and standard error). *Source* Authors' elaboration of World Bank indicators

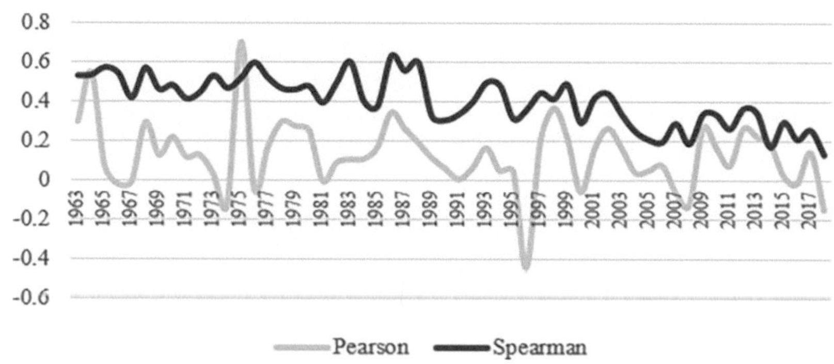

Fig. 7.2 Trends over time in parametric (Pearson) and nonparametric (Spearman) correlation coefficients within the (−1) lagged time series of the per cent share of agricultural land in total landscape. *Source* Authors' elaboration of World Bank indicators

with several countries experiencing land stock increases and many other undergoing a net cropland loss. Multiple drivers underlying such diversified regime seem to be active in both affluent and emerging economies.

As a typical example, Puerto Rico has lost cropland because of intense industrialization and tourism expansion. On the contrary, countries as Saudi Arabia, Cuba and Uganda, have greatly increased the percentage of agricultural land. For instance, Saudi Arabia experienced a sharp expansion in high value farmland thanks to the joint impact of multiple agricultural policies and reforms, inducing a non-negligible production of wheat for export purposes. Another completely different example is Italy, a country that has experienced a typical dynamic of industrialized economies with a sharp drop in the percentage of agricultural land (from 70.3 to 41.7%) in connection with a strong decrease in the number of farm holders.

Multidimensional Analysis

Principal Component Analysis (PCA) was finally adopted to explore the complex structure of the input data, considering a two-way matrix with countries and years.

Overall, the first component (PC1) captures nearly all the input variance (94.8%), indicating strong (global) trends in agricultural land-use. PC2 adds some minor (but possibly non-negligible) secondary variations (4%), likely reflecting regional (or country-specific) influences. The remaining components contribute marginally, implying that most differences in agricultural land-use are well-explained by just the first two components. PC1 basically reflects broad patterns in agricultural land-use, including the geography of land suitability for agriculture, as well as long-term soil practices and rural developmental policies. Climatic conditions added to this complex framework. PC2 likely captures localized factors reflected in the intrinsic difference in agricultural land-use intensity across world regions and countries, variations in land-use policies, economic priorities, and population density or growth.

Figure 7.3 illustrates the main results of Principal Component Analysis (PCA) projecting countries and years onto a biplot that uses Components 1 and 2 as axes. Years projected near the origin of the axes indicate minimal variability and delineate stable periods of agricultural land-use for most countries. Longer vectors suggest greater temporal contributions to the overall variance in the input data, potentially driven by global or regional changes, as economic shifts, land use reforms, or climate-related policies. Countries tightly clustered near the origin, e.g. Paraguay (PRY), Bermuda (BMU), and Guam (GUM), exhibit average values of agricultural land stock, with minimal deviation from the overall trend captured by the first two dimensions.

Outliers such as Saudi Arabia (SAU), Puerto Rico (PRI), and Uruguay (URI), exhibit peculiar characteristics in terms of agricultural land use. Saudi Arabia lays on the far negative side of Component 1, indicating an extremely low percentage of agricultural land relative to its total

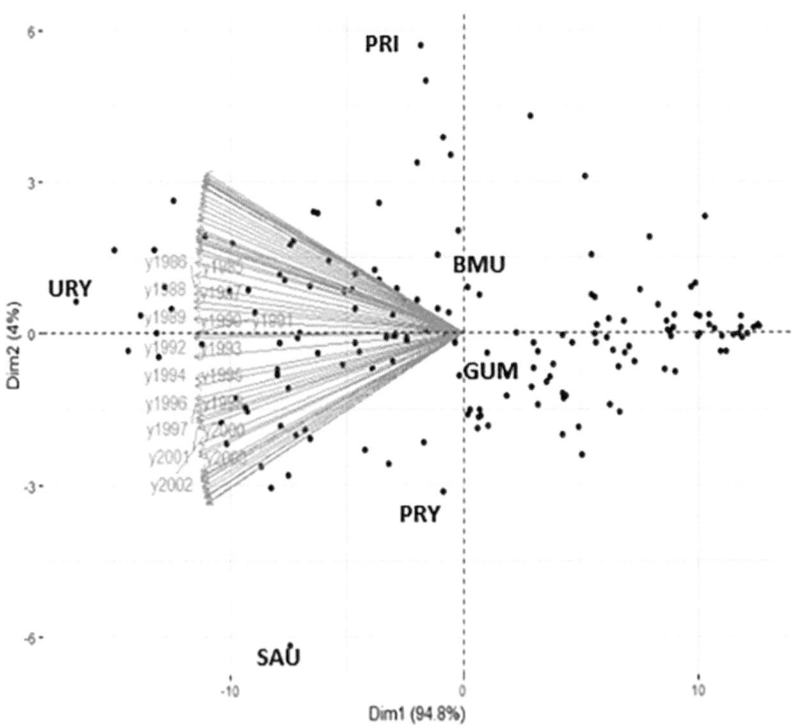

Fig. 7.3 Results of a principal component analysis illustrated in a biplot that reflects the distribution of variables' (i.e. years) loadings and rows' (i.e. countries) scores along the first two axes. *Source* Authors' elaboration of World Bank indicators

area, coherent with its environment constituted by arid land, limiting the availability of arable land, heavily relying on imports for food provision. Puerto Rico is located far on the positive side of Component 2, likely for the limited total land area disproportionately affecting agricultural land stocks. Finally, Uruguay positioned along the positive side of Component 1, depicting a very high percentage of agricultural land relative to its area, as typical of an agrarian economy extensively using land for livestock grazing and crop production.

Discussion

Homogeneous agricultural land stocks—as revealed in the results of different statistics over a sufficient long time and across a particularly vast sample of world countries—reflect the importance of the primary sector, irrespective of its modest contribution to global value added chains (e.g. Salvati & Zitti, 2009). However, results also indicate some countries having significantly lost an evident stock of agricultural land. Islands such as Cyprus and Malta, as well as small agglomerations of Caribbean islands, and the American and British Virgin Islands, have lost the largest part of cultivated lands, being replaced with a service-based economic development, mainly grounded on tourism expansion (Ferrara et al., 2017; Lamonica et al., 2020; Salvati & Carlucci, 2014).

For instance, after more than four centuries of an agricultural-dominated economic system in Puerto Rico (e.g. Lewis, 1955), the twentieth century saw unexpected sociopolitical changes resulting in a sudden shift to industry (Giusti-Cordero, 1996). Coupled with population growth, structural changes resulted in urbanization (Sallustio et al., 2018; Salvati et al., 2012; Tomao et al., 2017). The same dynamics were observed in several affluent countries (Egidi et al., 2020; Seifollahi-Aghmiuni et al., 2022; Vardopoulos et al., 2021). In Italy, from the 1960s onwards, industrialization shifting population to urban areas, caused the sudden decrease of agricultural enterprises, and, in turn, a sharp decrease in farmland stocks (Ferrara et al., 2016; Salvati & Carlucci, 2011; Salvati & Zitti, 2008).

Examples of a reverse trend can be also mentioned. For instance, the primary sector of Saudi Arabia has gradually gained importance, accounting for 13% of Gross Domestic Product (GDP) by 1985, in turn felling to 7% in 1990 and 3% in 2014. Cuba, Nigeria, Paraguay, and Burundi also experienced a sudden increase in agricultural land, although with a less direct impact on value added chains (Yang & Solangi, 2024). The important role of land grabbing—a particularly relevant phenomenon in some countries—in shaping agricultural land stocks all over the world should be better focussed in specific studies integrating multiple information sources.

Concluding Remarks

Taken as a leading sustainability indicator, agricultural land stocks per country provide effective information driving governmental decisions in developing action plans in favour of a 'complete' sustainability of agricultural productions (Anderson et al., 2017). Food security and environmental protection–aimed at the achievement of some of the SDGs ultimately favouring zero hunger and zero poverty contexts–are the main issues at stake when using this indicator for sustainability assessment (Soergel et al., 2021), especially in the light of a better knowledge of local communities and regional dynamics (Wilson et al., 2017). A specific, country-level analysis may highlight the environmental impact of agricultural production contributing to implement concrete actions of land planning (Streimikis & Baležentis, 2020), promoting food safety, social advancements, and economic growth (Francaviglia et al., 2019). While quantitative analyses usually neglect the impact of Covid–19 on agricultural land stocks and the related activities, longer series of data, including the recent pandemic history, should be considered, and additional information from different sources (satellite imagery, field surveys, land cadastres) should be collected to precisely assess such dynamics.

References

Agarwala, C., Jemaneh, J., & Kassie, Y. (2022). Government policies and sustainable food systems: Navigating challenges, seizing opportunities, and advancing environmental and social resilience. *Law and Economics, 16*(2), 88–102.

Amundson, R., Berhe, A. A., Hopmans, J. W., Olson, C., Sztein, A. E., & Sparks, D. L. (2015). Soil and human security in the 21st century. *Science, 348*(6235), 1261071.

Anderson, K., Ryan, B., Sonntag, W., Kavvada, A., & Friedl, L. (2017). Earth observation in service of the 2030 Agenda for Sustainable Development. *Geo-Spatial Information Science, 20*(2), 77–96.

Atukunda, P., Eide, W. B., Kardel, K. R., Iversen, P. O., & Westerberg, A. C. (2021). Unlocking the potential for achievement of the UN Sustainable Development Goal 2–'Zero Hunger'–in Africa: Targets, strategies, synergies and challenges. *Food & Nutrition Research, 65*, 10–29219.

Aydinalp, C., & Cresser, M. S. (2008). The effects of global climate change on agriculture. *American-Eurasian Journal of Agricultural & Environmental Sciences, 3*(5), 672–676.

Baldos, U. L. C., & Hertel, T. W. (2014). Global food security in 2050: The role of agricultural productivity and climate change. *Australian Journal of Agricultural and Resource Economics, 58*(4), 554–570.

Bello, M. M., Yahaya, J. U., & Adamu, I. (2024). An analysis of sustainable agricultural productivity and food security in Nigeria. *Journal of Political Discourse, 2*(1), 2.

Berhane, A. (2018). Climate change and variability impacts on agricultural productivity and food security. *Climate Weather Forecasting, 6*(240), 2.

Bindi, M., & Olesen, J. E. (2011). The responses of agriculture in Europe to climate change. *Regional Environmental Change, 11*(S1), 151–158.

Bizier, V., Gennari, P., & Kalamvrezos Navarro, D. (2022). Role of international, regional and country organizations in adapting to statistical standards and regional differences: The case of food and agriculture statistics. *Statistical Journal of the IAOS, 38*(2), 511–532.

Bolarinwa, O. D., Ogundari, K., & Aromolaran, A. B. (2020). Intertemporal evaluation of household food security and its determinants: Evidence from Rwanda. *Food Security, 12*(1), 179–189.

Bununu, Y. A., Bello, A., & Ahmed, A. (2023). Land cover, land use, climate change and food security. *Sustainable Earth Reviews, 6*(1), 16.

Campbell, B. M., Vermeulen, S. J., Aggarwal, P. K., Corner-Dolloff, C., Girvetz, E., Loboguerrero, A. M., & Wollenberg, E. (2016). Reducing risks to food security from climate change. *Global Food Security, 11*, 34–43.

Chapman, M., Walker, W. S., Cook-Patton, S. C., Ellis, P. W., Farina, M., Griscom, B. W., & Baccini, A. (2020). Large climate mitigation potential from adding trees to agricultural lands. *Global Change Biology, 26*(8), 4357–4365.

Chen, X., Shuai, C., Gao, J., & Wu, Y. (2023). Analyzing the socioeconomic determinants of PM2. 5 air pollution at the global level. *Environmental Science and Pollution Research, 30*(10), 27257–27269.

Chen, X., Shuai, C., & Wu, Y. (2023a). Global food stability and its socio-economic determinants towards sustainable development goal 2 (Zero Hunger). *Sustainable Development, 31*(3), 1768–1780.

Dar, W. D., & Laxmipathi Gowda, C. L. (2013). Declining agricultural productivity and global food security. *Journal of Crop Improvement, 27*(2), 242–254.

Dorward, A., Kydd, J., Morrison, J., & Urey, I. (2004). A policy agenda for pro-poor agricultural growth. *World Development, 32*(1), 73–89.

Egidi, G., Salvati, L., & Vinci, S. (2020). The long way to tipperary: City size and worldwide urban population trends, 1950–2030. *Sustainable Cities and Society, 60*, Article 102148.

Eichhorn, M. P., Paris, P., Herzog, F., Incoll, L. D., Liagre, F., Mantzanas, K., & Dupraz, C. (2006). Silvoarable systems in Europe–past, present and future prospects. *Agroforestry Systems, 67*(1), 29–50.

Elliott, J., Deryng, D., Müller, C., Frieler, K., Konzmann, M., Gerten, D., & Wisser, D. (2014). Constraints and potentials of future irrigation water availability on agricultural production under climate change. *Proceedings of the National Academy of Sciences, 111*(9), 3239–3244.

Emina, K. A. (2021). Sustainable development and the future generations. *Social Sciences, Humanities and Education Journal, 2*(1), 57–71.

Ferrara, A., Kelly, C., Wilson, G. A., Nolè, A., Mancino, G., Bajocco, S., & Salvati, L. (2016). Shaping the role of 'fast' and 'slow' drivers of change in forest-shrubland socio-ecological systems. *Journal of Environmental Management, 169*, 155–166.

Ferrara, C., Carlucci, M., Grigoriadis, E., Corona, P., & Salvati, L. (2017). A comprehensive insight into the geography of forest cover in Italy: Exploring the importance of socioeconomic local contexts. *Forest Policy and Economics, 75*, 12–22.

Francaviglia, R., Di Bene, C., Farina, R., Salvati, L., & Vicente-Vicente, J. L. (2019). Assessing "4 per 1000" soil organic carbon storage rates under Mediterranean climate: A comprehensive data analysis. *Mitigation and Adaptation Strategies for Global Change, 24*, 795–818.

Fritz, S., See, L., & Rembold, F. (2010). Comparison of global and regional land cover maps with statistical information for the agricultural domain in Africa. *International Journal of Remote Sensing, 31*(9), 2237–2256.

Gebremedhin, H., Gebresamual, G., Abadi, N., Hailemariam, M., Teka, K., & Mesfin, S. (2018). Conversion of communal grazing land into arable land and its impacts on soil properties and vegetation cover. *Arid Land Research and Management, 32*(2), 236–252.

Gennari, P., Seck, P., Guerrero, M., Toma, I., & Shama, R. (2022). The FAO-UN Women partnership to support countries in producing sex-disaggregated data on agricultural landownership. *Statistical Journal of the IAOS, 38*(3), 893–914.

Giusti-Cordero, J. A. (1996). Labour, ecology and history in a Puerto Rican plantation region: 'classic' rural proletarians revisited. *International Review of Social History, 41*(S4), 53–82.

Gornall, J., Betts, R., Burke, E., Clark, R., Camp, J., Willett, K., & Wiltshire, A. (2010). Implications of climate change for agricultural productivity in the early twenty-first century. *Philosophical Transactions of the Royal Society B: Biological Sciences, 365*(1554), 2973–2989.

Grote, U. (2014). Can we improve global food security? A socio-economic and political perspective. *Food Security, 6*(2), 187–200.

Hamidov, A., Helming, K., & Balla, D. (2016). Impact of agricultural land use in Central Asia: A review. *Agronomy for Sustainable Development, 36*(1), 6.

Heltberg, R., Siegel, P. B., & Jorgensen, S. L. (2009). Addressing human vulnerability to climate change: Toward a 'no-regrets' approach. *Global Environmental Change, 19*(1), 89–99.

Henson, S., & Jaffee, S. (2006). Food safety standards and trade: Enhancing competitiveness and avoiding exclusion of developing countries. *The European Journal of Development Research, 18*(4), 593–621.

Hertel, T. W., & Lobell, D. B. (2014). Agricultural adaptation to climate change in rich and poor countries: Current modeling practice and potential for empirical contributions. *Energy Economics, 46*, 562–575.

Humphreys, S. (2022). Against future generations. *European Journal of International Law, 33*(4), 1061–1092.

Incerti, G., Feoli, E., Salvati, L., Brunetti, A., & Giovacchini, A. (2007). Analysis of bioclimatic time series and their neural network-based classification to characterise drought risk patterns in South Italy. *International Journal of Biometeorology, 51*, 253–263.

Ingutia, R., & Sumelius, J. (2022). Determinants of food security status with reference to women farmers in rural Kenya. *Scientific African, 15*, Article e01114.

Jiang, L., Deng, X., & Seto, K. C. (2013). The impact of urban expansion on agricultural land use intensity in China. *Land Use Policy, 35*, 33–39.

Kanter, D. R., Schwoob, M. H., Baethgen, W. E., Bervejillo, J. E., Carriquiry, M., Dobermann, A., & de Lima, J. M. S. (2016). Translating the sustainable development goals into action: A participatory backcasting approach for developing national agricultural transformation pathways. *Global Food Security, 10*, 71–79.

Knickel, K., Redman, M., Darnhofer, I., Ashkenazy, A., Chebach, T. C., Šūmane, S., & Rogge, E. (2018). Between aspirations and reality: Making farming, food systems and rural areas more resilient, sustainable and equitable. *Journal of Rural Studies, 59*, 197–210.

Kumar, P., Tokas, J., Kumar, N., Lal, M., & Singal, H. R. (2018). Climate change consequences and its impact on agriculture and food security. *International Journal of Chemical Studies, 6*(6), 124–133.

Lamonica, G. R., Recchioni, M. C., Chelli, F. M., & Salvati, L. (2020). The efficiency of the cross-entropy method when estimating the technical coefficients of input–output tables. *Spatial Economic Analysis, 15*(1), 62–91.

Langhelle, O. (1999). Sustainable development: Exploring the ethics of our common future. *International Political Science Review, 20*(2), 129–149.

Lewis, G. K. (1955). Puerto Rico: A case-study of change in an underdeveloped area. *The Journal of Politics, 17*(4), 614–650.

Luan, Y., Cui, X., & Ferrat, M. (2013). Historical trends of food self-sufficiency in Africa. *Food Security, 5*(3), 393–405.

Mancino, G., Nolè, A., Salvati, L., & Ferrara, A. (2016). In-between forest expansion and cropland decline: A revised USLE model for soil erosion risk under land-use change in a Mediterranean region. *Ecological Indicators, 71*, 544–550.

Mârza, B., Angelescu, C., & Tindeche, C. (2015). Agricultural insurances and food security. The new climate change challenges. *Procedia Economics and Finance, 27*, 594–599.

Movilla-Pateiro, L., Mahou-Lago, X. M., Doval, M. I., & Simal-Gandara, J. (2021). Toward a sustainable metric and indicators for the goal of sustainability in agricultural and food production. *Critical Reviews in Food Science and Nutrition, 61*(7), 1108–1129.

Moyer, J. D., & Hedden, S. (2020). Are we on the right path to achieve the sustainable development goals? *World Development, 127*, Article 104749.

Mugambiwa, S. S., & Tirivangasi, H. M. (2017). Climate change: A threat towards achieving 'Sustainable Development Goal number two' (end hunger, achieve food security and improved nutrition and promote sustainable agriculture) in South Africa. *Jàmbá: Journal of Disaster Risk Studies, 9*(1), 1–6.

Mulungh, M. G. (2021). Impact of climate change on biodiversity and food security: A global perspective—A review article. *Agriculture and Food Security, 10*(1), 1–25.

Nastis, S. A., Michailidis, A., & Chatzitheodoridis, F. (2012). Climate change and agricultural productivity. *African Journal of Agricultural Research, 7*(35), 4885–4893.

Natarajan, N., Newsham, A., Rigg, J., & Suhardiman, D. (2022). A sustainable livelihoods framework for the 21st century. *World Development, 155*, Article 105898.

Palmisano, G. O., Govindan, K., Boggia, A., Loisi, R. V., De Boni, A., & Roma, R. (2016). Local Action groups and rural sustainable development. A spatial multiple criteria approach for efficient territorial planning. *Land Use Policy, 59*, 12–26.

Pandey, B., & Seto, K. C. (2015). Urbanization and agricultural land loss in India: Comparing satellite estimates with census data. *Journal of Environmental Management, 148*, 53–66.

Parry, M., Rosenzweig, C., Iglesias, A., Fischer, G., & Livermore, M. (1999). Climate change and world food security: A new assessment. *Global Environmental Change, 9*, S51–S67.

Persson, A. S., Olsson, O., Rundlöf, M., & Smith, H. G. (2010). Land use intensity and landscape complexity—Analysis of landscape characteristics in an agricultural region in Southern Sweden. *Agriculture, Ecosystems and Environment, 136*(1–2), 169–176.

Peter, V. H., & Peng, Y. A. N. G. (2014). How could agricultural land systems contribute to raise food production under global change? *Journal of Integrative Agriculture, 13*(7), 1432–1442.

Pham, L. V., & Smith, C. (2014). Drivers of agricultural sustainability in developing countries: A review. *Environment Systems and Decisions, 34*(2), 326–341.

Pinstrup-Andersen, P. (2002). Food and agricultural policy for a globalizing world: Preparing for the future. *American Journal of Agricultural Economics, 84*(5), 1201–1214.

Piorr, H. P. (2003). Environmental policy, agri-environmental indicators and landscape indicators. *Agriculture, Ecosystems and Environment, 98*(1–3), 17–33.

Pongratz, J., Reick, C., Raddatz, T., & Claussen, M. (2008). A reconstruction of global agricultural areas and land cover for the last millennium. *Global Biogeochemical Cycles, 22*(3).

Prabhakar, S. V. R. K. (2021). A succinct review and analysis of drivers and impacts of agricultural land transformations in Asia. *Land Use Policy, 102*, Article 105238.

Pretty, J. N., Morison, J. I., & Hine, R. E. (2003). Reducing food poverty by increasing agricultural sustainability in developing countries. *Agriculture, Ecosystems and Environment, 95*(1), 217–234.

Qi, X., Fu, Y., Wang, R. Y., Ng, C. N., Dang, H., & He, Y. (2018). Improving the sustainability of agricultural land use: An integrated framework for the conflict between food security and environmental deterioration. *Applied Geography, 90*, 214–223.

Quintas-Soriano, C., Castro, A. J., Castro, H., & García-Llorente, M. (2016). Impacts of land use change on ecosystem services and implications for human well-being in Spanish drylands. *Land Use Policy, 54*, 534–548.

Ramankutty, N., Evan, A. T., Monfreda, C., & Foley, J. A. (2008). Farming the planet: 1. Geographic distribution of global agricultural lands in the year 2000. *Global Biogeochemical Cycles, 22*(1).

Rasul, G. (2016). Managing the food, water, and energy nexus for achieving the sustainable development goals in South Asia. *Environmental Development, 18*, 14–25.

Rhodes, C. J. (2014). Soil erosion, climate change and global food security: Challenges and strategies. *Science Progress, 97*(2), 97–153.

Ruane, J., & Sonnino, A. (2011). Agricultural biotechnologies in developing countries and their possible contribution to food security. *Journal of Biotechnology, 156*(4), 356–363.

Sallustio, L., Pettenella, D., Merlini, P., Romano, R., Salvati, L., Marchetti, M., & Corona, P. (2018). Assessing the economic marginality of agricultural lands in Italy to support land use planning. *Land Use Policy, 76*, 526–534.

Salvati, L., & Carlucci, M. (2011). The economic and environmental performances of rural districts in Italy: Are competitiveness and sustainability compatible targets? *Ecological Economics, 70*(12), 2446–2453.

Salvati, L., & Carlucci, M. (2014). A composite index of sustainable development at the local scale: Italy as a case study. *Ecological Indicators, 43*, 162–171.

Salvati, L., Perini, L., Sabbi, A., & Bajocco, S. (2012). Climate aridity and land use changes: A regional-scale analysis. *Geographical Research, 50*(2), 193–203.

Salvati, L., & Zitti, M. (2008). Regional convergence of environmental variables: Empirical evidences from land degradation. *Ecological Economics, 68*(1–2), 162–168.

Salvati, L., & Zitti, M. (2009). Substitutability and weighting of ecological and economic indicators: Exploring the importance of various components of a synthetic index. *Ecological Economics, 68*(4), 1093–1099.

Seifollahi-Aghmiuni, S., Kalantari, Z., Egidi, G., Gaburova, L., & Salvati, L. (2022). Urbanisation-driven land degradation and socioeconomic challenges in peri-urban areas: Insights from Southern Europe. *Ambio, 51*(6), 1446–1458.

Slater, R., Peskett, L., Ludi, E., & Brown, D. (2007). Climate change, agricultural policy and poverty reduction—How much do we know. *Natural Resource Perspectives, 109*(7), 1–6.

Soergel, B., Kriegler, E., Weindl, I., Rauner, S., Dirnaichner, A., Ruhe, C., & Popp, A. (2021). A sustainable development pathway for climate action within the UN 2030 Agenda. *Nature Climate Change, 11*(8), 656–664.

Song, X. P., Hansen, M. C., Stehman, S. V., Potapov, P. V., Tyukavina, A., Vermote, E. F., & Townshend, J. R. (2018). Global land change from 1982 to 2016. *Nature, 560*(7720), 639–643.

Spiertz, J. H. J. (2009). Nitrogen, sustainable agriculture and food security: A review. *Sustainable Agriculture*, 635–651.

Strassburg, B. B., Latawiec, A. E., Barioni, L. G., Nobre, C. A., Da Silva, V. P., Valentim, J. F., & Assad, E. D. (2014). When enough should be enough: Improving the use of current agricultural lands could meet production demands and spare natural habitats in Brazil. *Global Environmental Change, 28*, 84–97.

Streimikis, J., & Baležentis, T. (2020). Agricultural sustainability assessment framework integrating sustainable development goals and interlinked priorities of environmental, climate and agriculture policies. *Sustainable Development, 28*(6), 1702–1712.

Swaminathan, M. S., & Bhavani, R. V. (2013). Food production and availability-Essential prerequisites for sustainable food security. *Indian Journal of Medical Research, 138*(3), 383–391.

Tomao, A., Quatrini, V., Corona, P., Ferrara, A., Lafortezza, R., & Salvati, L. (2017). Resilient landscapes in

Mediterranean urban areas: Understanding factors influencing forest trends. *Environmental Research, 156*, 1–9.

Tscharntke, T., Clough, Y., Wanger, T. C., Jackson, L., Motzke, I., Perfecto, I., & Whitbread, A. (2012). Global food security, biodiversity conservation and the future of agricultural intensification. *Biological Conservation, 151*(1), 53–59.

Van Veen, T. W. S. (2005). International trade and food safety in developing countries. *Food Control, 16*(6), 491–496.

Vardopoulos, I., Tsilika, E., Sarantakou, E., Zorpas, A. A., Salvati, L., & Tsartas, P. (2021). An integrated SWOT-PESTLE-AHP model assessing sustainability in adaptive reuse projects. *Applied Sciences, 11*(15), 7134.

Volkov, S. N., & Fomin, A. A. (2018). Provision of effective regulation of land relation and efficient agricultural land use. *International Agricultural Journal, 3*, 64–71.

Weiss, E. B. (1990). In fairness to future generations. *Environment: Science and Policy for Sustainable Development, 32*(3), 6–31.

Wheeler, T., & Von Braun, J. (2013). Climate change impacts on global food security. *Science, 341*(6145), 508–513.

Wilson, G. A., Kelly, C. L., Briassoulis, H., Ferrara, A., Quaranta, G., Salvia, R., Detsis, V., Curfs, M., Cerda, A., El-Aich, A., Liu, H., Kosmas, C., Alados, C. L., Imeson, A., Landgrebe-Trinkunaite, R., Salvati, L., Naumann, S., Danwen, H., Iosifides, T., … Zhang, P. (2017). Social memory and the resilience of communities affected by land degradation. *Land Degradation and Development, 28*(2), 383–400.

Yang, T. M., Lo, J., & Shiang, J. (2015). To open or not to open? Determinants of open government data. *Journal of Information Science, 41*(5), 596–612.

Yang, Z., & Solangi, Y. A. (2024). Analyzing the relationship between natural resource management, environmental protection, and agricultural economics for sustainable development in China. *Journal of Cleaner Production, 450*, Article 141862.

Zhang, X., & Cai, X. (2011). Climate change impacts on global agricultural land availability. *Environmental Research Letters, 6*(1), Article 014014.

Zhen, L., & Routray, J. K. (2003). Operational indicators for measuring agricultural sustainability in developing countries. *Environmental Management, 32*(1), 34–46.

8 Forest Rents

Introduction

By providing shelter to almost 1.6 billion people, forests have acquired a crucial role in guaranteeing vital services, both regulatory and cultural (e.g. spiritual, recreational and aesthetic) ones (Newbold et al., 2016; Newton et al., 2020; Zahawi et al., 2013). They preserve most of terrestrial biodiversity and significantly contain the damages from natural disasters such as floods, droughts, and landslides (Bonjour et al., 2013; Mancino et al., 2016; Vardopoulos et al., 2021). Moreover, forests contribute to the balance of oxygen, carbon dioxide, in turn protecting watersheds, which provide around 75% of the world freshwater (Kauppi et al., 2018; Salvati & Zitti, 2009; Tomao et al., 2017). Forests also stabilize ecosystems, promoting sustainable agriculture and food security (Aerts & Honnay, 2011; Cumming et al., 2014; Wali et al., 2017). Sustainable projects, dealing with renewable energy and biofuels, foster inclusive access to resources thus creating several job opportunities and green industries (Chandratreya, 2025; Falcone, 2023; Oloyede et al., 2024); this benefits rural areas by generating stable income (Kaygusuz, 2012). In addition, it empowers women in forest management and energy (Batra, 2023; Salvati & Carlucci, 2011; Seifollahi-Aghmiuni et al., 2022).

An effective transition towards renewable energy also fosters skill development, thus strengthening local economies, job market and social resilience at large (Akinsemolu & Onyeaka, 2025; Daoudi, 2024; Peskett et al., 2007). The reallocation of time, previously employed in fuel collection, towards sustainable projects, allows effective actions to attain some crucial development goals, as those connected to education, equity and gender equality (Olabi et al., 2023). Within this healthier framework, cleaner energy sources should contain indoor air pollution, thus reduce respiratory illnesses and enhance food security, nutrition, and water quality (Kumar et al., 2023; Rosenthal et al., 2018; Volodzkiene, 2025). Innovative and efficient technologies reduce environmental impact and encourage industrial development (Mata et al., 2021). Investments in renewable energy infrastructures should address energy gaps, benefiting sectors like health care, education and transportation, while innovations in forest management promote sustainability (Majid, 2020).

Effective policies towards sustainable environment could also result in an enhanced financial system (Birdsall & Wheeler, 1993) and an improved efficiency in economic transactions for technological progress providing environmental benefits (Lamonica et al., 2020; Levine, 1997). Ramos-Meza et al. (2021) analyse the development of a green financial system envisaging the achievement of Millennium Development Goals. To this purpose, an efficient system to contain carbon emissions should discourage environmental degradation (e.g. Adewumi et al., 2024;

C. Ciaschini and L. Salvati, *Sustainable Development*, Sustainable Development Goals Series,
https://doi.org/10.1007/978-3-032-12559-0_8

Dasgupta et al., 2001; Islam et al., 2013). In general, sustainable forest and energy practices address multiple SDGs, driving environmental, economic and social progress, while promoting global cooperation (Ferrara et al., 2017; Mancino et al., 2016; Sallustio et al., 2018). A sustainable management of forests can improve water quality (SDG6), reduce emissions (SDG13) and support biodiversity (SDG15). The interconnections among environmental, social and economic goals encourage policymakers to design more cohesive and effective development strategies (Adanma & Ogunbiyi, 2024; Ferrara et al., 2016; Wilson et al., 2017).

A straightforward access to renewable energy supports health (SDG3), gender equality (SDG5), and poverty reduction (SDG1), providing cumulative benefits in favour of resilient societies, containing natural disasters, and preserving productive agricultural systems (e.g. De Groot et al., 2002; Dhakal & Kattel, 2019; Munang et al., 2013; Murphy, 2012). The relevant contribution of forests in supplying energy services triggers a progressive transition towards a low-carbon and circular development relying on (renewable) natural resources, i.e. the bio-economy (Halonen et al., 2022; Ladu et al., 2020; Marchi et al., 2018). To this aim, products derived from healthy forest continuously preserve diversified forms of life on land (Anikwe & Ife, 2023; Pandey et al., 2016; Sheppard et al., 2020). All these efforts should ensure long-term benefits to land-based ecosystems (e.g. Incerti et al., 2007; Egidi et al., 2020; Gogoi et al., 2020).

Literature Review

Forest rents acquire a crucial relevance under the economic and climate viewpoints in regional and local contexts where forests constitute a natural and cultural heritage of invaluable value (e.g. Agnoletti et al., 2022). Analysing the economic and climatic determinants of forest rents in percentage of Gross Domestic Product (GDP) allows to understand not only the role of the forest sector in the national economy, but also its contribution to environmental sustainability and the containment of climate changes (Da Silva et al., 2023; Nkem et al., 2010; Perez-Garcia et al. 1997). Forest rents define the economic benefits deriving from a sustainable management of wood resources (Kant, 2004). These benefits include the production of wood, paper, derivate products, but also ecosystem services as the safeguard of soil, conservation of biodiversity and rural tourism (Patterson & Coelho, 2009; Pearce, 2001; Siry et al., 2005).

For instance, the forest sector in a possibly advanced (while internally divided) country such as Italy constitutes about 1%-2% of national GDP, highlighting a strategic sector for the safeguard of natural heritage and the development of sustainable economic activities (Di Cori et al., 2022; Grilli et al., 2015; Gutiérrez & Lozano, 2022). Under this economic viewpoint, forest rents are affected by several factors: (i) market trends in forest products' prices, (ii) policies of sustainable land management, (iii) the international demand for wood and related products, as well as (iv) the ability to enhance resources through tourism and recreational activities (Willemen et al., 2013). Sustainable forest management, indeed, can generate significant rents, contributing to economic expansion of rural and mountain areas (Egoh et al., 2012).

Nevertheless, it is important to highlight that the explicit quantification of forest rents as a per cent share in the Gross Domestic Product (GDP) of a given country could vary over time, following conservation policies, investments in the primary sector, and market dynamics (Moiseyev et al., 2011). In addition, forest rents define an important flow of revenues for national, regional and local administrative authorities, through taxes, licences and other regulation provisions (Lanfredi et al., 2023). An effective management of these resources can favour a sustainable economic development, create employment and promote practices of sustainable forestry (Buongiorno, 1996). The economic valorisation of forests, however, should go along with policies of safeguard and preservation ensuring long-term sustainability, avoiding the risk of excessive (basically, unsustainable) exploitation and, in turn, depletion of natural resources (e.g. Agnoletti &

Santoro, 2018; Chen & Innes, 2013; Gebre & Gebremedhin, 2019).

Under the climate viewpoint, forests acquire a crucial role in mitigating climate changes thanks to their ability of absorbing carbon dioxide from the atmosphere and to store it in the wood and in soils (Zahoor et al., 2022). Healthy and well-managed forests contribute to the reduction of greenhouse emissions (Belcher, 2005), representing a key element in the contrast of climate changes at global levels (Arslan et al., 2022). Forest rents, therefore, are not only an economic benefit, but also a valuable ecosystem service to safeguard climate (Crook & Clapp, 1998; Kumar et al., 2024; Lindner et al., 2010).

In sustainably managed forests, a higher capacity of CO_2 absorption should correspond to contained negative effects of climate changes (Sharma & Henriques, 2005). Nevertheless, global warming and climate changes themselves represent a direct threat to old forests (Calfapietra et al., 2015). Namely, extreme meteorological events, wildfires, droughts and pests are compromising the health and absorption capacity of forests (Wastl et al., 2012). Within this context, forest rents acquire a double role. On the one side, they constitute an economic incentive to a safe and sustainable management of resources; on the other side, they represent a natural capital to be preserved for the purposes of climate change mitigations (Allen et al., 2010). The valorisation of forest as carbon sinks can be pursued implementing policies of environmental compensation, thus strengthening the connection between climate and economic aspects (Bateman & Mace, 2020; Lavalle et al., 2009; Neumayer, 1998).

Forests and Climate Change

In recent times, climate change represents one of the most compelling challenges, affecting significantly natural resources and global economies (Sivakumar, 2011). Among its multiple components, the forest sector acquires a crucial role, both as source of greenhouse gas emissions and as key element to the mitigation and adaptation of climate risks (Bretschger & Pittel, 2020; Ravindranath, 2007; Stern, 2008). Economic literature estimated the contribution of forest rents to GDP and their level of exposure to some climatic and political factors (Hertel & Rosch, 2010). Several studies highlighted that global warming and extreme climate phenomena have direct and indirect effects on forests (Aaheim et al., 2011). Specifically, the Intergovernmental Panel on Climate Change (IPCC) research highlights that the increase in temperatures, together with events as fires, droughts and floods, is likely to reduce forest resources, driving a relevant loss of biodiversity and negatively affecting the dynamics of growth of plants (Badr & El-Shazly, 2024; Rawat et al., 2024; Thanh et al., 2024). These occurrences also affect the forest productivity, with direct consequences generated by the sector (Wunder, 2005).

In macroeconomic terms, forest rents contribute significantly to GDP of many nations, in particular those with economies relying on natural resources (Fu & Liu, 2023). According to the literature, the share of forest rents on GDP can notably vary, affected by factors as resources' management, environmental policies, and climate change (Lin & Cai, 2023). Several empirical studies analysed the impact of climate change on forest rents (Qin et al., 2022). Literature agrees on the fact that the negative impact of climate on forest resources results in a decrease of forest rents and, in turn, in their per cent share in total GDP (Repetto, 1987; Barbier and Burgess, 1997; Fisher et al., 2011). In regions exposed to frequent wildfires, the loss of forests and wood production implies a relevant decrease of rents, in some cases until 20%-30% with respect to earlier values (Delacote, 2007).

Some studies, however, suggest that in some areas, the adaptation through techniques of sustainable forest management and investment in innovative technologies can mitigate these effects, stabilizing (or even increasing) the share of forest rents in total GDP (Andrée et al., 2019). This strategy is not universally applicable and requires effective policies and long-term investments (Corbera et al., 2010). Projections based on climate models indicate that persisting high levels of greenhouse gas emissions could deteriorate forest environment, with a consequent, further decrease in forest rents and their share in total GDP (Peng et al., 2023). This could induce ripple effects on the economy and on subsistence means of rural communities (Povellato et al., 2007).

Academic literature also provides suggestions for mitigation policies, e.g. implementing sustainable practices and promoting adaptation technologies, considered as crucial to preserve sector's profitability (Keenan, 2015). Climate change negatively impacted production and forest rents, affecting their share in total GDP (Perez-Garcia et al., 2002). The adaptation capacity and sustainable management policies are key tools to contain its negative effects (Berman et al., 2012; Huang et al., 2024; Smit & Pilifosova, 2003). Nevertheless, the future of forest sector and its contribution to economy will rely on the speed and effectiveness of the actions implemented to promote sustainable practices (Mbow et al., 2014). Climate change represents one of the main challenges for the world economies, affecting forestry in significant ways (Locatelli et al., 2015). The economic literature has extensively focused on the role of forest resources in a context of climate change (Bhattarai & Hammig, 2001).

Earlier studies highlighted the impact of climate change on the distribution and productivity of forests, increasing the frequency and intensity of wildfires, pests and extreme meteorological events (Hao et al., 2019; Pramova et al., 2012; Spies et al., 2010). These phenomena bring to a reduction in forest resources affecting their ability of contributing to economic dynamics (Vijaya Venkata Raman et al. 2012). The per cent share of forest rents in GDP varies significantly among countries and regions. According to some projections, in economic systems dependent from forestry (some countries in Africa and South America), forest resources can contribute to 2–5% of GDP, while this percentage tends to be often lower than 1% in more developed economies (Mamo et al., 2007). Climate warming may lower productivity and the economic value of forest resources, affecting their economic contribution to GDP (Sloan & Sayer, 2015). Within this context, sustainable forest management and mitigation policies should aim at preserving and potentially enhancing the economic role of forests (Schoene & Bernier, 2012).

Forests and Economic Dynamics

Forest rents are defined as surplus values—net of all costs and returns—commonly adopted to quantify natural capital. In recent years, almost all countries, particularly in developing contexts, experienced market volatility (Apergis & Payne, 2014; Guan et al., 2020; Omri & Kahouli, 2014). These circumstances, strengthened by the constant pursuit of economic growth, led to the discovery of a wide number of natural resources (Shahbaz et al., 2019). Differently to developed countries, within these emerging contexts, relevant issues arise on how to effectively manage these resources, involving efficient and sustainable policies (Cavalcanti et al., 2011; Khan et al., 2020; Satti et al., 2014).

Nevertheless, affluent countries have often experienced the 'resource curse' (or 'paradox of plenty'), failing in fully benefiting from their natural resources provision (Borge et al., 2015; Leibbrandt & Lynham, 2018; Moti, 2019). Governments of these countries, indeed, often adopt authoritarian policies, likely to cause internal conflicts, at the expense of economic growth (Sachs & Warner, 2001). The 'resource curse' prevents an effective transition to green development for two main reasons. First, economies heavily relying on natural resource revenues experience higher levels of rent-seeking and corruption, thus discouraging an effective implementation of environmental regulations (Cheng et al., 2021; Li & Tong, 2024; Puscaciu & Puscaciu, 2015). Moreover, a significant income from natural resources' rent hinders the transition towards environmentally sustainable development and growth (Alpha & Ding, 2016; Esu, 2017; Yu, 2023).

Nevertheless, empirical evidence also show negative, even not significant relationships, when studying the linkage between political stability, corruption and economic growth. The protection of natural resources crucially supports a proper economic and financial development, increasing productivity and, in turn, income. Over the years, revenues from natural resources acquired importance; more recently, topic concerning the ecological degradation fostered the debate within scholars. Although GDP positively affects pollution, clean energy could also produce negative externalities (Zhang & Zhang, 2023). On the one side, it contributes to technological progress, job creation and knowledge improvement (Zhang & Dilanchiev, 2022). On the other side, a disproportionate resource exploitation could cause detrimental environmental consequences as pollution and climate change, threatening public health and sustainable development—also boosted by human action—through deforestation or mining (Han et al., 2022).

Recent literature analysed the connection between forest rents and economic growth. Studies concerning Sub-Saharan African and Asian economies also adopt econometric models to highlight a positive impact of forest rents on economic growth (Huang et al., 2020; Jović et al., 2016; Ofori & Grechyna, 2021; Xie et al., 2022). Chernor Momodu and Unisa (2022) and Kwakwa et al. (2022) show similar outcomes for Tunisia's economy. In addition, Satti et al. (2014) analysed the connection between natural resources and economic development in Venezuela between 1971 and 2011. Financial growth relying on an excessive extraction of natural resources could cause the degradation of natural ecosystems—seriously challenging the long-term environmental health (Liu et al., 2022). The use of natural resources relevantly impacts financial and economic development, among others, of tourism sector, universally recognized as the fastest growing business (Auty, 2000). Although tourism relevantly contributes to financial development, the wide use of energy and fossil fuels for transportation and lodging leads to negative environmental consequences (Taghizadeh-Hesary et al., 2022).

Empirical Findings

The spatial distribution of forest rents in total GDP was studied for 1970 and 2021 at the global scale. In general, the wealth of forests crucially affects all countries, particularly African regions and Asiatic tropical areas, thus constituting a key component of local economies. At the beginning of the study period, a more imperative economic role of forests was observed in tropical and subtropical countries rather than in tempered countries. This reflects a higher level of biodiversity and wider size of forest area in regions historically exploited for natural resources. In 1970, economies of different African, Asian and south American countries were significantly connected with forests and forestry, while industrialized countries performed a lower dependence from these resources. Data from year 2021 confirm the importance of forest rents in the different countries, particularly Central and Southern Africa, as well as tropical Asia. As in the previous case, industrialized countries have exhibited a less pronounced level of dependence from forest rents.

Descriptive Statistics

The most relevant information on the statistical distribution of forest rents across world countries are displayed in Table 8.1 at regular intervals over time (1970, 1980, 1990, 2000, 2010, and 2021). In general, metrics (descriptive statistics) shape highly asymmetric, heavy tailed, distributions of forest rents for all years, characterized by a tendency towards extremely high values. Median values, constantly lower than the mean value, indicate asymmetry in forest rents across world countries. Although the arithmetic mean of forest rents fluctuates between 1.6 and 3.3%, it was clearly decreasing over time, with the lower value (1.6%) recorded in 2021. Asymmetry depends on outliers; a kurtosis significantly higher than 3 (with a peak of 17.1 in 1980) delineates distributions with heavier tails, i.e. a higher probability of extreme observations for individual forest rents at the country scale.

Table 8.1 Descriptive statistics representing the distribution of the per cent share of forest rents in Gross Domestic Product (GDP) across world countries

Metric	1970	1980	1990	2000	2010	2021
Mean	2.1	3.3	2.7	2.0	2.0	1.6
Minimum	0.0	0.0	0.0	0.0	0.0	0.0
Maximum	16.6	44.6	21.0	15.1	23.7	18.4
Normal range	7.8	13.5	7.8	7.4	11.7	11.6
Median	0.6	0.8	0.5	0.3	0.4	0.2
25 percentile	0.1	0.1	0.1	0.0	0.0	0.0
75 percentile	2.9	3.3	3.8	2.7	2.3	1.9
Skewness	2.1	3.9	2.2	2.0	3.2	3.1
Kurtosis	4.8	17.1	4.6	3.5	11.7	11.1

Source Authors' elaboration of World Bank indicators

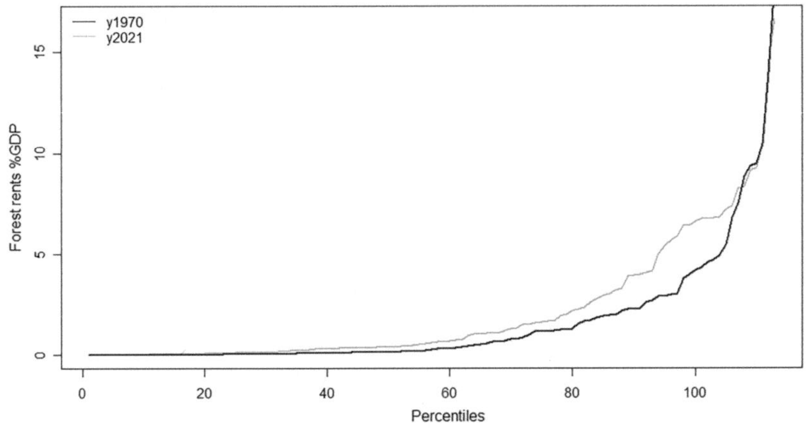

Fig. 8.1 Graphical summary of the statistical distribution of forest rents across world countries as a percentage of GDP, 1970 and 2021, based on a percentile plot. *Source* Authors' elaboration of World Bank indicators

Outliers, observed for each decade, indicate a strong dependence of some countries from forest resources, diminishing in number and intensity over time. After 1990, a rather limited number of outliers and the decreasing interquartile range remark a slight reduction in forest rents as percentage of GDP, thus reflecting a persisting transition towards economies less dependent on forest resources. This tendency should be connected with a higher degree of economic diversification implying a more sustainable exploitation of forests. Figure 8.1 shows the evolution of forest rents as a percentage of GDP in two different years, 1970 and 2021, through a percentile distribution. The plot suggests a global decrease between 1970 and 2021 for most countries, likely because of structural changes in global economies. In particular, economic diversification and the decline in forest resources in some countries could, in turn, imply a different use of natural capital. A marked polarization seems to occur in 1970 (grey line), when some countries were highly dependent on forest resources; this dependence decreased in 2021, with a gradual increase in few countries for the last percentiles (black line).

Correlation Analysis

Figure 8.2 comparatively illustrates trends over time in Spearman (non-parametric, nonlinear) and Pearson (parametric, linear) correlation coefficients with the aim at delineating possible breakpoints in the observed time series. Both correlation measures exhibit high values (close to 1), suggesting a strong relationship between the variables over time, namely a year-by-year homogeneity in the statistical distribution of forest rents across world countries (e.g. Salvati & Zitti,

2008). In particular, Pearson correlation shows high values throughout the whole period with negligible variations across the years. Conversely, Spearman correlation highlights occasional deviations from the constant trend over time, possibly connected to outliers, less detectable considering Pearson measures.

Multidimensional Analysis

The outcomes of a Principal Component Analysis (PCA) run on input variables indicate the dominance of the first two axes when explaining the overall matrix variance. In particular, Component 1 (PC1) accounted for 76.7% of the total variance; Component 2 (PC2) explained an additional 12.7% of variance, providing a comprehensive summary of input information when considered together. Based on the first two components, the biplot illustrated in Fig. 8.3 displays the latent relationship between countries (black dots) and years (grey arrows). Most countries were tightly clustered around the origin of the axes, implying similar values of forest rents. Nevertheless, some countries, i.e. Burundi, Democratic Republic of Congo, Guinea-Bissau, Equatorial Guinea, New Caledonia, and Uganda, are located far from this central cluster. Countries distant from such a cluster (i.e. Guinea-Bissau, Democratic Republic of Congo, and New Caledonia) could represent cases with consistently higher (or lower) dependence on forest rents differing from the general trend. The grey arrows projecting the study years (e.g. 1991, 2000) highlight the directions and magnitude of each year's contribution to principal components.

Most years, located on the right side of the plot, indicate a high correlation among forest rent values across these years and consistent forest rent trends over time for many countries. In particular, the indicator seems to show a steady pattern over time, perhaps affected by stable forest management or economic factors common to many countries. Years slightly distant from this cluster (e.g. earlier years such as 1971, or specific years in the subsequent decades) might represent time periods when forest rents deviated more significantly from a standard trend, probably due to policy changes, economic shifts, or environmental events (e.g. Salvati & Carlucci, 2014). In general, the biplot highlights a rather stable economic trend in forest rents, with a notable homogeneity among years of the observed time series. This trend probably reflects the influence of global economic factors, as the increasing demand of natural resources and human pressure (Francaviglia et al., 2019). The observations far from the centre of the graph could reflect countries that, for economic or environmental reasons, differ from the average trend, as those highly relying on forests or characterized by unique national policies.

Discussion

The safeguard of forests and sustainable management of their resources has been a relevant concern of policymakers at a global level (Cubbage et al., 2007). This topic has been discussed in several international meetings as the XIV World Congress on Forests held in Durban in 2015 (MacDicken et al., 2015). In that circumstance, the relevance of new collaborations between agricultural, finance and energy sectors has been remarked in a perspective of sustainable development of forests (Baumgartner, 2019; Hahn & Knoke, 2010; Swamy et al., 2018). As already mentioned, this issue has also been included among the targets of the 2030 Development Agenda of United Nations, requiring relevant and innovative financial commitments supporting forest environments (Akhtar-Schuster et al., 2011).

To this purpose, the present work focuses on the role and trend of forest rents aimed at economic development (Arvanitis & Weigert, 2017). In general, the study highlights a rather stable trend in forest rents, with a notable homogeneity across years, probably connected to global economic factors, as the increasing demand of natural resources and the pressure for a sustainable management of forests (Banskota et al., 2014; Burnley, 2011; Nitoslawski et al., 2021). Countries as Burundi, Democratic Republic of Congo, Uganda and Guinea-Bissau seem to have a more pronounced forest rent dependence. These

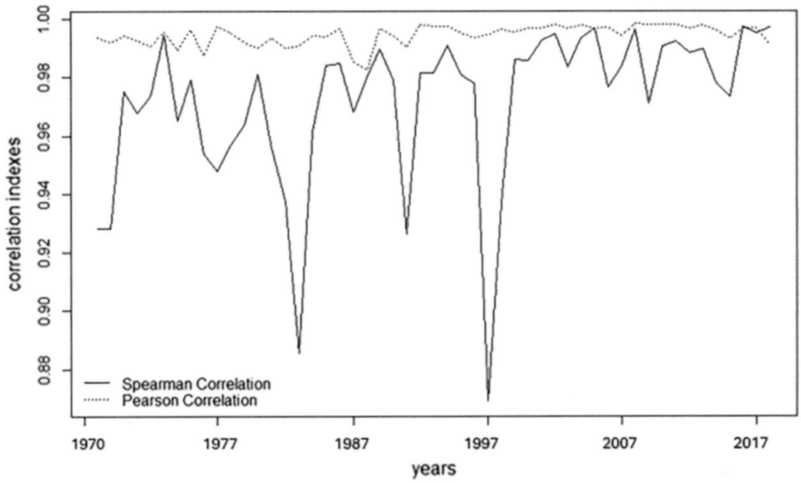

Fig. 8.2 Trends over time in parametric (Pearson) and non-parametric (Spearman) correlation coefficients within the (−1) lagged time series of forest rents over GDP. *Source* Authors' elaboration of World Bank indicators

nations could be significantly reliant on forestry for economic growth or revenue generation, highlighting the major role of forestry activities in their economies (Wilkie et al., 2002). In particular, Burundi and Democratic Republic of Congo experienced higher ecological or economic impacts from forest resource use as well as greater challenges related to deforestation, biodiversity or governance sustainability issues (Levin et al., 2008; Chichinye et al., 2024; Mukaku Kazadi et al., 2024).

In recent times, several theoretical and empirical studies attempt to catch these general trends, analysing long-term transformations in forest coverage and the related economic factors (Austin, 2008; Clark et al., 2018; Rjoub et al., 2021). Forest management, afforestation and reforestation should also consider the relative change in values of agricultural lands with respect

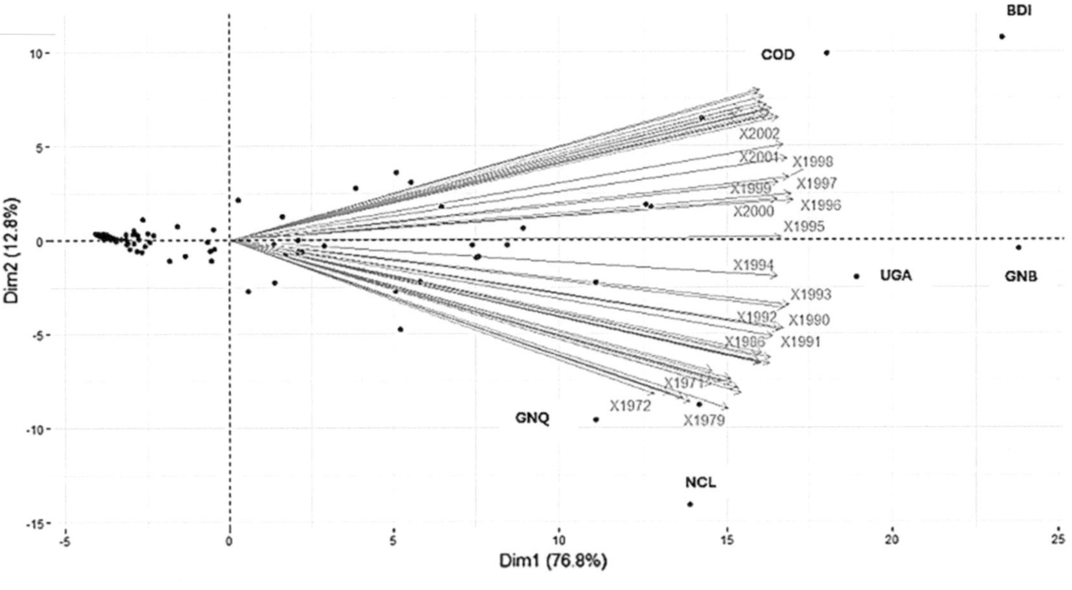

Fig. 8.3 Results of a principal component analysis as illustrated in a biplot reflecting the distribution of variables' (i.e. years) loadings and rows (i.e. countries) scores along the first two axes. *Source* Authors' elaboration of World Bank indicators

to forests (Hurmekoski & Hetemäki, 2013; Rudel et al., 2005; Salvati et al., 2012; Tegegne et al., 2019). The extent of policies in favour of sustainable forest management as the United Nations programme for the Reduction of the Emissions from Deforestation and forest Degradation (REDD +) should be designed accounting for the level of soil exploitation of each country (Dieng et al., 2023; Dulal et al., 2012; Miah, 2020).

Concluding Remarks

Countries with high forest cover and low percentage of deforestation should design policies to improve forest management, while those with a high level of deforestation should focus on projects of afforestation and reduction of deforestation processes. Future research should rely on a wider and updated database to monitor sudden transformations in forest environments. This should also include economic factors and their correlations with policies and investments to preserve forests.

References

Aaheim, A., Gopalakrishnan, R., Chaturvedi, R. K., Ravindranath, N. H., Sagadevan, A. D., Sharma, N., & Wei, T. (2011). A macroeconomic analysis of adaptation to climate change impacts on forests in India. *Mitigation and Adaptation Strategies for Global Change, 16*, 229–245.

Adanma, U. M., & Ogunbiyi, E. O. (2024). A comparative review of global environmental policies for promoting sustainable development and economic growth. *International Journal of Applied Research in Social Sciences, 6*(5), 954–977.

Adewumi, A., Olu-lawal, K. A., Okoli, C. E., Usman, F. O., & Usiagu, G. S. (2024). Sustainable energy solutions and climate change: A policy review of emerging trends and global responses. *World Journal of Advanced Research and Reviews, 21*(2), 408–420.

Aerts, R., & Honnay, O. (2011). Forest restoration, biodiversity and ecosystem functioning. *BMC Ecology, 11*, 1–10.

Agnoletti, M., Piras, F., Venturi, M., & Santoro, A. (2022). Cultural values and forest dynamics: The Italian forests in the last 150 years. *Forest Ecology and Management, 503*, Article 119655.

Agnoletti, M., & Santoro, A. (2018). Rural landscape planning and forest management in Tuscany (Italy). *Forests, 9*(8), 473.

Akhtar-Schuster, M., Thomas, R. J., Stringer, L. C., Chasek, P., & Seely, M. (2011). Improving the enabling environment to combat land degradation: Institutional, financial, legal and science-policy challenges and solutions. *Land Degradation and Development, 22*(2), 299–312.

Akinsemolu, A. A., & Onyeaka, H. (2025). The role of green education in achieving the sustainable development goals: A review. *Renewable and Sustainable Energy Reviews, 210*, Article 115239.

Allen, C. D., Macalady, A. K., Chenchouni, H., Bachelet, D., McDowell, N., Vennetier, M., & Cobb, N. (2010). A global overview of drought and heat-induced tree mortality reveals emerging climate change risks for forests. *Forest Ecology and Management, 259*(4), 660–684.

Alpha, B. B., & Ding, Y. (2016). A study on the impact of natural resources endowment on economic growth: Empirical evidence from Mali. *Journal of Economics and Development Studies, 4*(4), 81–103.

Andrée, B. P. J., Chamorro, A., Spencer, P., Koomen, E., & Dogo, H. (2019). Revisiting the relation between economic growth and the environment; a global assessment of deforestation, pollution and carbon emission. *Renewable and Sustainable Energy Reviews, 114*, Article 109221.

Anikwe, M. A. N., & Ife, K. (2023). The role of soil ecosystem services in the circular bioeconomy. *Frontiers in Soil Science, 3*, 1209100.

Apergis, N., & Payne, J. E. (2014). The oil curse, institutional quality, and growth in MENA countries: Evidence from time-varying cointegration. *Energy Economics, 46*, 1–9.

Arslan, H. M., Khan, I., Latif, M. I., Komal, B., & Chen, S. (2022). Understanding the dynamics of natural resources rents, environmental sustainability, and sustainable economic growth: New insights from China. *Environmental Science and Pollution Research, 29*(39), 58746–58761.

Arvanitis, Y., & Weigert, M. (2017). Turning resource curse into development dividends in Guinea-Bissau. *Resources Policy, 53*, 226–237.

Austin, G. (2008). Resources, techniques, and strategies south of the Sahara: Revising the factor endowments perspective on African economic development, 1500–2000 1. *The Economic History Review, 61*(3), 587–624.

Auty, R. M. (2000). How natural resources affect economic development. *Development Policy Review, 18*(4).

Badr, Ų. A., & El-Shazly, H. H. (2024). Climate change and biodiversity loss: interconnected challenges and priority measures. *Catrina: The International Journal of Environmental Sciences, 29*(1), 69–78.

Banskota, A., Kayastha, N., Falkowski, M. J., Wulder, M. A., Froese, R. E., & White, J. C. (2014). Forest monitoring using Landsat time series data: A review. *Canadian Journal of Remote Sensing, 40*(5), 362–384.

Barbier, E. B., & Burgess, J. C. (1997). The economics of tropical forest land use options. *Land Economics*, 174–195.

Bateman, I. J., & Mace, G. M. (2020). The natural capital framework for sustainably efficient and equitable decision making. *Nature Sustainability, 3*(10), 776–783.

Batra, G. (2023). Renewable energy economics: Achieving harmony between environmental protection and economic goals. *Social Science Chronicle, 2*(2), 1–32.

Baumgartner, R. J. (2019). Sustainable development goals and the forest sector—A complex relationship. *Forests, 10*(2), 152.

Belcher, B. M. (2005). Forest product markets, forests and poverty reduction. *International Forestry Review, 7*(2), 82–89.

Berman, R., Quinn, C., & Paavola, J. (2012). The role of institutions in the transformation of coping capacity to sustainable adaptive capacity. *Environmental Development, 2*, 86–100.

Bhattarai, M., & Hammig, M. (2001). Institutions and the environmental Kuznets curve for deforestation: a crosscountry analysis for Latin America, Africa and Asia. *World development, 29*(6), 995–1010.

Birdsall, N., & Wheeler, D. (1993). Trade policy and industrial pollution in Latin America: Where are the pollution havens? *The Journal of Environment and Development, 2*(1), 137–149.

Bonjour, S., Adair-Rohani, H., Wolf, J., et al. (2013). Solid fuel use for household cooking: Country and regional estimates for 1980–2010. *Environmental Health Perspectives, 121*(7, 784–90.

Borge, L. E., Parmer, P., & Torvik, R. (2015). Local natural resource curse? *Journal of Public Economics, 131*, 101–114.

Bretschger, L., & Pittel, K. (2020). Twenty key challenges in environmental and resource economics. *Environmental and Resource Economics, 77*, 725–750.

Buongiorno, J. (1996). Forest sector modeling: A synthesis of econometrics, mathematical programming, and system dynamics methods. *International Journal of Forecasting, 12*(3), 329–343.

Burnley, C. (2011). Natural resources conflict in the Democratic Republic of the Congo: A question of governance. *Sustainable Development Law & Policy, 12*, 7.

Calfapietra, C., Barbati, A., Perugini, L., Ferrari, B., Guidolotti, G., Quatrini, A., & Corona, P. (2015). Carbon mitigation potential of different forest ecosystems under climate change and various managements in Italy. *Ecosystem Health and Sustainability, 1*(8), 1–9.

Cavalcanti, T. V. D. V., Mohaddes, K., & Raissi, M. (2011). Growth, development and natural resources: New evidence using a heterogeneous panel analysis. *The Quarterly Review of Economics and Finance, 51*(4), 305–318.

Chandratreya, A. (2025). The role of renewable energy in promoting social equity and inclusive economic growth. *Renewable Energy and the Economic Welfare of Society*, 183–212.

Chen, J., & Innes, J. L. (2013). The implications of new forest tenure reforms and forestry property markets for sustainable forest management and forest certification in China. *Journal of Environmental Management, 129*, 206–215.

Cheng, Z., Li, X., & Wang, M. (2021). Resource curse and green economic growth. *Resources Policy, 74*, Article 102325.

Chernor Momodu, B., & Unisa, D. (2022). Economic growth effects of the interaction of natural resources and institutional quality by source: Empirical evidence from Africa. *International Journal of Developing and Emerging Economies, 10*(2), 1–33.

Chichinye, A., Mlambo, D., Sebata, A., & Mabidi, A. (2024). Policy regarding sustainable forest management and resources. *Agroforestry for Carbon and Ecosystem Management*, 403–417.

Clark, R., Reed, J., & Sunderland, T. (2018). Bridging funding gaps for climate and sustainable development: Pitfalls, progress and potential of private finance. *Land Use Policy, 71*, 335–346.

Corbera, E., Estrada, M., & Brown, K. (2010). Reducing greenhouse gas emissions from deforestation and forest degradation in developing countries: Revisiting the assumptions. *Climatic Change, 100*(3), 355–388.

Crook, C., & Clapp, R. A. (1998). Is market-oriented forest conservation a contradiction in terms? *Environmental Conservation, 25*(2), 131–145.

Cubbage, F., Harou, P., & Sills, E. (2007). Policy instruments to enhance multi-functional forest management. *Forest Policy and Economics, 9*(7), 833–851.

Cumming, G. S., Buerkert, A., Hoffmann, E. M., Schlecht, E., von Cramon-Taubadel, S., & Tscharntke, T. (2014). Implications of agricultural transitions and urbanization for ecosystem services. *Nature, 515*(7525), 50–57.

Da Silva, J. G., de Almeida, R. B., & Carvalho, L. V. (2023). An economic analysis of a zero-deforestation policy in the Brazilian Amazon. *Ecological Economics, 203*, Article 107613.

Daoudi, M. (2024). Education in renewable energies: A key factor of Morocco's 2030 energy transition project. Exploring the impact on SDGs and future perspectives. *Social Sciences and Humanities Open, 9*, 100833.

Dasgupta, S., Laplante, B., & Mamingi, N. (2001). Pollution and capital markets in developing countries. *Journal of Environmental Economics and Management, 42*(3), 310–335.

De Groot, R. S., Wilson, M. A., & Boumans, R. M. (2002). A typology for the classification, description and valuation of ecosystem functions, goods and services. *Ecological Economics, 41*(3), 393–408.

Delacote, P. (2007). Agricultural expansion, forest products as safety nets, and deforestation. *Environment and Development Economics, 12*(2), 235–249.

Dhakal, B., & Kattel, R. R. (2019). Effects of global changes on ecosystems services of multiple natural resources in mountain agricultural landscapes. *Science of the Total Environment, 676*, 665–682.

Di Cori, V., Robert, N., Franceschinis, C., Pettenella, D. M., & Thiene, M. (2022). Framework proposal to quantify the contribution of non-wood forest products to the European Union forest-based bioeconomy. *Forests, 13*(3), 362.

Dieng, M., Mbow, C., Skole, D. L., & Ba, B. (2023). Sustainable land management policy to address land degradation: Linking old forest management practices in Senegal with new REDD+ requirements. *Frontiers in Environmental Science, 11*, 1088726.

Dulal, H. B., Shah, K. U., & Sapkota, C. (2012). Reducing emissions from deforestation and forest degradation (REDD) projects: Lessons for future policy design and implementation. *International Journal of Sustainable Development and World Ecology, 19*(2), 116–129.

Egidi, G., Salvati, L., & Vinci, S. (2020). The long way to tipperary: City size and worldwide urban population trends, 1950–2030. *Sustainable Cities and Society, 60*, Article 102148.

Egoh, B. N., O'Farrell, P. J., Charef, A., Gurney, L. J., Koellner, T., Abi, H. N., & Willemen, L. (2012). An African account of ecosystem service provision: Use, threats and policy options for sustainable livelihoods. *Ecosystem Services, 2*, 71–81.

Esu, G. E. (2017). Assessing the resource curse question: A case of crude oil production in Nigeria. *Journal of Economic Research, 22*(2), 153–213.

Falcone, P. M. (2023). Sustainable energy policies in developing countries: A review of challenges and opportunities. *Energies, 16*(18), 6682.

Ferrara, A., Kelly, C., Wilson, G. A., Nolè, A., Mancino, G., Bajocco, S., & Salvati, L. (2016). Shaping the role of 'fast' and 'slow' drivers of change in forest-shrubland socio-ecological systems. *Journal of Environmental Management, 169*, 155–166.

Ferrara, C., Carlucci, M., Grigoriadis, E., Corona, P., & Salvati, L. (2017). A comprehensive insight into the geography of forest cover in Italy: Exploring the importance of socioeconomic local contexts. *Forest Policy and Economics, 75*, 12–22.

Fisher, B., Lewis, S. L., Burgess, N. D., Malimbwi, R. E., Munishi, P. K., Swetnam, R. D., & Balmford, A. (2011). Implementation and opportunity costs of reducing deforestation and forest degradation in Tanzania. *Nature Climate Change, 1*(3), 161–164.

Francaviglia, R., Di Bene, C., Farina, R., Salvati, L., & Vicente-Vicente, J. L. (2019). Assessing "4 per 1000" soil organic carbon storage rates under Mediterranean climate: A comprehensive data analysis. *Mitigation and Adaptation Strategies for Global Change, 24*, 795–818.

Fu, R., & Liu, J. (2023). Revenue sources of natural resources rents and its impact on sustainable development: Evidence from global data. *Resources Policy, 80*, Article 103226.

Gebre, T., & Gebremedhin, B. (2019). The mutual benefits of promoting rural-urban interdependence through linked ecosystem services. *Global Ecology and Conservation, 20*, Article e00707.

Gogoi, B., Nath, T., Kashyap, D., Sarma, S., & Kalita, R. (2020). Sustainable agriculture, forestry and fishery for bioeconomy. In *Current developments in Biotechnology and Bioengineering* (pp. 349–371). Elsevier.

Grilli, G., Nikodinoska, N., Paletto, A., & De Meo, I. (2015). Stakeholders' preferences and economic value of forest ecosystem services: An example in the Italian Alps. *Baltic Forestry, 21*(2), 298–307.

Guan, J., Kirikkaleli, D., Bibi, A., & Zhang, W. (2020). Natural resources rents nexus with financial development in the presence of globalization: Is the "resource curse" exist or myth? *Resources Policy, 66*, Article 101641.

Gutiérrez, E., & Lozano, S. (2022). Cross-country comparison of the efficiency of the European forest sector and second stage DEA approach. *Annals of Operations Research, 314*(2), 471–496.

Hahn, W. A., & Knoke, T. (2010). Sustainable development and sustainable forestry: Analogies, differences, and the role of flexibility. *European Journal of Forest Research, 129*, 787–801.

Halonen, M., Näyhä, A., & Kuhmonen, I. (2022). Regional sustainability transition through forest-based bioeconomy? Development actors' perspectives on related policies, power, and justice. *Forest Policy and Economics, 142*, Article 102775.

Han, Y., Tan, S., Zhu, C., & Liu, Y. (2022). Research on the emission reduction effects of carbon trading mechanism on power industry: plant-level evidence from China.

Hao, Y., Xu, Y., Zhang, J., Hu, X., Huang, J., Chang, C. P., & Guo, Y. (2019). Relationship between forest resources and economic growth: Empirical evidence from China. *Journal of Cleaner Production, 214*, 848–859.

Hertel, T. W., & Rosch, S. D. (2010). Climate change, agriculture, and poverty. *Applied Economic Perspectives and Policy, 32*(3), 355–385.

Huang, D., Shen, H., Miao, Y., Ding, R., Lin, Y., & Tan, H. (2024). The impacts of forest resources, green investment, healthcare, and education on environmental pollution: China Carbon neutrality program. *Journal of Cleaner Production, 467*, Article 143038.

Huang, Y., Raza, S. M. F., Hanif, I., Alharthi, M., Abbas, Q., & Zain-ul-Abidin, S. (2020). The role of forest resources, mineral resources, and oil extraction in economic progress of developing Asian economies. *Resources Policy, 69*, Article 101878.

Hurmekoski, E., & Hetemäki, L. (2013). Studying the future of the forest sector: Review and implications for long-term outlook studies. *Forest Policy and Economics, 34*, 17–29.

Incerti, G., Feoli, E., Salvati, L., Brunetti, A., & Giovacchini, A. (2007). Analysis of bioclimatic time series and their neural network-based classification to characterise drought risk patterns in South Italy. *International Journal of Biometeorology, 51*, 253–263.

International Journal of Climate Change Strategies and Management, 15(2), 212–231.

Islam, F., Shahbaz, M., Ahmed, A. U., & Alam, M. M. (2013). Financial development and energy consumption nexus in Malaysia: A multivariate time series analysis. *Economic Modelling, 30*, 435–441.

Jović, S., Maksimović, G., & Jovović, D. (2016). Appraisal of natural resources rents and economic development. *Resources Policy, 50*, 289–291.

Kant, S. (2004). Economics of sustainable forest management. *Forest Policy and Economics, 6*(3), 197.

Kauppi, P. E., Sandström, V., & Lipponen, A. (2018). Forest resources of nations in relation to human well-being. *PLoS ONE, 13*(5), Article e0196248.

Kaygusuz, K. (2012). Energy for sustainable development: A case of developing countries. *Renewable and Sustainable Energy Reviews, 16*(2), 1116–1126.

Keenan, R. J. (2015). Climate change impacts and adaptation in forest management: A review. *Annals of Forest Science, 72*, 145–167.

Khan, Z., Hussain, M., Shahbaz, M., Yang, S., & Jiao, Z. (2020). Natural resource abundance, technological innovation, and human capital nexus with financial development: A case study of China. *Resources Policy, 65*, Article 101585.

Kumar, A., Shabnam, S., Oraon, P. R., & Malik, M. S. (2024). Forest Health in a Changing Scenario of Climate Change. In *Sustainable Forest Resources Management*, 109–129.

Kumar, P., Singh, A. B., Arora, T., Singh, S., & Singh, R. (2023). Critical review on emerging health effects associated with the indoor air quality and its sustainable management. *Science of the Total Environment, 872*, Article 162163.

Kwakwa, P. A., Adzawla, W., Alhassan, H., & Achaamah, A. (2022). Natural resources and economic growth: Does political regime matter for Tunisia? *Journal of Public Affairs, 22*, Article e2707.

Ladu, L., Imbert, E., Quitzow, R., & Morone, P. (2020). The role of the policy mix in the transition toward a circular forest bioeconomy. *Forest Policy and Economics, 110*, Article 101937.

Lamonica, G. R., Recchioni, M. C., Chelli, F. M., & Salvati, L. (2020). The efficiency of the cross-entropy method when estimating the technical coefficients of input–output tables. *Spatial Economic Analysis, 15*(1), 62–91.

Lanfredi, M., Coluzzi, R., Imbrenda, V., Nosova, B., Giacalone, M., Turco, R., & Salvati, L. (2023). In-between environmental sustainability and economic viability: An analysis of the state, regulations, and future of Italian forestry sector. *Land, 12*(5), 1001.

Lavalle, C., Micale, F., Houston, T. D., Camia, A., Hiederer, R., Lazar, C., & Genovese, G. (2009). Climate change in Europe. 3. Impact on agriculture and forestry. A review. *Agronomy for Sustainable Development*, *29*, 433–446.

Leibbrandt, A., & Lynham, J. (2018). Does the paradox of plenty exist? Experimental evidence on the curse of resource abundance. *Experimental Economics, 21*, 337–354.

Levin, K., McDermott, C., & Cashore, B. (2008). The climate regime as global forest governance: Can reduced emissions from Deforestation and Forest Degradation (REDD) initiatives pass a dual effectiveness test? *International Forestry Review, 10*(3), 538–549.

Levine, R. (1997). Financial development and economic growth: Views and agenda. *Journal of Economic Literature, 35*(2), 688–726.

Li, X., & Tong, X. (2024). Fostering green growth in Asian developing economies: The role of good governance in mitigating the resource curse. *Resources Policy, 90*, Article 104724.

Lin, C., & Cai, P. (2023). Analyzing the impacts of natural resource utilization and green economic growth in China: Evidence from a econometric analysis. *Resources Policy, 81*, Article 103311.

Lindner, M., Maroschek, M., Netherer, S., Kremer, A., Barbati, A., Garcia-Gonzalo, J., & Marchetti, M. (2010). Climate change impacts, adaptive capacity, and vulnerability of European forest ecosystems. *Forest Ecology and Management, 259*(4), 698–709.

Liu, X., Tong, D., Huang, J., Zheng, W., Kong, M. & Zhou, G. (2022). What matters in the e-commerce era? Modelling and mapping shop rents in Guangzhou, China. *Land Use Policy*, 123, 106430.

Locatelli, B., Pavageau, C., Pramova, E., & Di Gregorio, M. (2015). Integrating climate change mitigation and adaptation in agriculture and forestry: Opportunities and trade-offs. *Wiley Interdisciplinary Reviews: Climate Change, 6*(6), 585–598.

MacDicken, K. G., Sola, P., Hall, J. E., Sabogal, C., Tadoum, M., & de Wasseige, C. (2015). Global progress toward sustainable forest management. *Forest Ecology and Management, 352*, 47–56.

Majid, M. (2020). Renewable energy for sustainable development in India: Current status, future prospects, challenges, employment, and investment opportunities. *Energy, Sustainability and Society, 10*(1), 1–36.

Mamo, G., Sjaastad, E., & Vedeld, P. (2007). Economic dependence on forest resources: A case from Dendi District, Ethiopia. *Forest policy and Economics*, *9*(8), 916–927.

Mancino, G., Nolè, A., Salvati, L., & Ferrara, A. (2016). In-between forest expansion and cropland decline: A revised USLE model for soil erosion risk under land-use change in a Mediterranean region. *Ecological Indicators, 71*, 544–550.

Marchi, E., Chung, W., Visser, R., Abbas, D., Nordfjell, T., Mederski, P. S., & Laschi, A. (2018). Sustainable forest operations (SFO): A new paradigm in a changing world and climate. *Science of the Total Environment, 634*, 1385–1397.

Mata, T. M., Oliveira, G. M., Monteiro, H., Silva, G. V., Caetano, N. S., & Martins, A. A. (2021). Indoor air quality improvement using nature-based solutions: Design proposals to greener cities. *International Journal of Environmental Research and Public Health, 18*(16), 8472.

Mbow, C., Smith, P., Skole, D., Duguma, L., & Bustamante, M. (2014). Achieving mitigation and adaptation to climate change through sustainable agroforestry practices in Africa. *Current Opinion in Environmental Sustainability, 6*, 8–14.

Miah, M. D. (2020). Reducing Emissions from deforestation and forest degradation (REDD+). *Life on Land*, 797–807.

Moiseyev, A., Solberg, B., Kallio, A. M. I., & Lindner, M. (2011). An economic analysis of the potential contribution of forest biomass to the EU RES target and its implications for the EU forest industries. *Journal of Forest Economics, 17*(2), 197–213.

Moti, U. G. (2019). Africa's natural resource wealth: A paradox of plenty and poverty. *Advances in Social Sciences Research Journal, 6*(7), 483–504.

Mukaku Kazadi, G., Mpanda Mukenza, M., Kikuni Tchowa, J., Malaisse, F., Kabongo Kabeya, C., Pitchou Meniko To Hulu, J. P., & Useni Sikuzani, Y. (2024). Assessment of spatial dynamics of forest cover in Lomami National Park (DR Congo), 2008–2024: Implications for conservation and sustainable ecosystem management. *Ecologies, 6*(1), 2.

Munang, R., Thiaw, I., Alverson, K., Liu, J., & Han, Z. (2013). The role of ecosystem services in climate change adaptation and disaster risk reduction. *Current Opinion in Environmental Sustainability, 5*(1), 47–52.

Murphy, K. (2012). The social pillar of sustainable development: a literature review and framework for policy analysis. *Sustainability: Science, practice and policy, 8*(1), 15–29.

Neumayer, E. (1998). Preserving natural capital in a world of uncertainty and scarce financial resources. *The International Journal of Sustainable Development & World Ecology, 5*(1), 27–42.

Newbold, T., Hudson, L. N., Arnell, A. P., Contu, S., De Palma, A., Ferrier, S., & Purvis, A. (2016). Has land use pushed terrestrial biodiversity beyond the planetary boundary? A global assessment. *Science, 353*(6296), 288–291.

Newton, P., Kinzer, A. T., Miller, D. C., Oldekop, J. A., & Agrawal, A. (2020). The number and spatial distribution of forest-proximate people globally. *One Earth, 3*(3), 363–370.

Nitoslawski, S. A., Wong-Stevens, K., Steenberg, J. W., Witherspoon, K., Nesbitt, L., & Konijnendijk Van Den Bosch, C. C. (2021). The digital forest: Mapping a decade of knowledge on technological applications for forest ecosystems. *Earth's Future, 9*(8), e2021EF002123.

Nkem, J., Kalame, F. B., Idinoba, M., Somorin, O. A., Ndoye, O., & Awono, A. (2010). Shaping forest safety nets with markets: Adaptation to climate change under changing roles of tropical forests in Congo Basin. *Environmental Science and Policy, 13*(6), 498–508.

Ofori, P. E., & Grechyna, D. (2021). Remittances, natural resource rent and economic growth in Sub-Saharan Africa. *Cogent Economics and Finance, 9*(1), 1979305.

Olabi, A. G., Abdelkareem, M. A., Mahmoud, M. S., Elsaid, K., Obaideen, K., Rezk, H., & Sayed, E. T. (2023). Green hydrogen: Pathways, roadmap, and role in achieving sustainable development goals. *Process Safety and Environmental Protection, 177*, 664–687.

Oloyede, C. T., Itabiyi, O. E., Popoola, O. A., Jekayinfa, S. O., Olaniyan, M. A., Adebayo, A. O., & Fattah, I. M. R. (2024). Navigating prospects and challenges for green fuels for achieving economic, environmental and ecological resilience: A scientific review. *Biofuels, 15*(7), 929–941.

Omri, A., & Kahouli, B. (2014). Causal relationships between energy consumption, foreign direct investment and economic growth: Fresh evidence from dynamic simultaneous-equations models. *Energy Policy, 67*, 913–922.

Pandey, A. K., Tripathi, Y. C., & Kumar, A. (2016). Non timber forest products (NTFPs) for sustained livelihood: Challenges and strategies. *Research Journal of Forestry, 10*(1), 1–7.

Patterson, T. M., & Coelho, D. L. (2009). Ecosystem services: Foundations, opportunities, and challenges for the forest products sector. *Forest Ecology and Management, 257*(8), 1637–1646.

Pearce, D. W. (2001). The economic value of forest ecosystems. *Ecosystem Health, 7*(4), 284–296.

Peng, L., Li, Y., Raza, S. A., & Shahzadi, I. (2023). Natural resources and environmental sustainability: COP26 targets from resources-based perspective. *Resources Policy, 83*, Article 103623.

Perez-Garcia, J., Joyce, L. A., Binkley, C. S., & McGuire, A. D. (1997). Economic impacts of climatic change on the global forest sector: An integrated ecological/economic assessment. *Critical Reviews in Environmental Science and Technology, 27*(S1), 123–138.

Perez-Garcia, J., Joyce, L. A., McGuire, A. D., & Xiao, X. (2002). Impacts of climate change on the global forest sector. *Climatic Change, 54*, 439–461.

Peskett, L., Slater, R., Stevens, C., & Dufey, A. (2007). Biofuels, agriculture and poverty reduction. *Natural Resource Perspectives, 107*, 1–6.

Povellato, A., Bosello, F., & Giupponi, C. (2007). Cost-effectiveness of greenhouse gases mitigation measures in the European agro-forestry sector: A literature survey. *Environmental Science and Policy, 10*(5), 474–490.

Pramova, E., Locatelli, B., Djoudi, H., & Somorin, O. A. (2012). Forests and trees for social adaptation to climate variability and change. *Wiley Interdisciplinary Reviews: Climate Change, 3*(6), 581–596.

Puscaciu, V., & Puscaciu, F. D. (2015). Paradox of plenty or resources curse? Knowledge Horizons.*Economics, 7*(3), 239.

Qin, Y., Xiao, X., Wigneron, J. P., Ciais, P., Canadell, J. G., Brandt, M., & Moore, B., III. (2022). Large loss and rapid recovery of vegetation cover and aboveground biomass over forest areas in Australia during 2019–2020. *Remote Sensing of Environment, 278*, Article 113087.

Ramos-Meza, C. S., Zhanbayev, R., Bilal, H., Sultan, M., Pekergin, Z. B., & Arslan, H. M. (2021). Does digitalization matter in green preferences in nexus of output volatility and environmental quality? *Environmental Science and Pollution Research, 28*, 66957–66967.

Ravindranath, N. H. (2007). Mitigation and adaptation synergy in forest sector. *Mitigation and Adaptation Strategies for Global Change, 12*, 843–853.

Rawat, A., Kumar, D., & Khati, B. S. (2024). A review on climate change impacts, models, and its consequences on different sectors: A systematic approach. *Journal of Water and Climate Change, 15*(1), 104–126.

Repetto, R. (1987). Economic incentives for sustainable production. *The Annals of Regional Science, 21*, 44–59.

Rjoub, H., Ifediora, C. U., Odugbesan, J. A., Iloka, B. C., Xavier Rita, J., Dantas, R. M., & Martins, J. M. (2021). Implications of governance, natural resources, and security threats on economic development: Evidence from Sub-Saharan Africa. *International Journal of Environmental Research and Public Health, 18*(12), 6236.

Rosenthal, J., Quinn, A., Grieshop, A. P., Pillarisetti, A., & Glass, R. I. (2018). Clean cooking and the SDGs: Integrated analytical approaches to guide energy interventions for health and environment goals. *Energy for Sustainable Development, 42*, 152–159.

Rudel, T. K., Coomes, O. T., Moran, E., Achard, F., Angelsen, A., Xu, J., & Lambin, E. (2005). Forest transitions: Towards a global understanding of land use change. *Global Environmental Change, 15*(1), 23–31.

Sachs, J. D., & Warner, A. M. (2001). The curse of natural resources. *European Economic Review, 45*(4–6), 827–838.

Sallustio, L., Pettenella, D., Merlini, P., Romano, R., Salvati, L., Marchetti, M., & Corona, P. (2018). Assessing the economic marginality of agricultural lands in Italy to support land use planning. *Land Use Policy, 76*, 526–534.

Salvati, L., & Carlucci, M. (2011). The economic and environmental performances of rural districts in Italy: Are competitiveness and sustainability compatible targets? *Ecological Economics, 70*(12), 2446–2453.

Salvati, L., & Carlucci, M. (2014). A composite index of sustainable development at the local scale: Italy as a case study. *Ecological Indicators, 43*, 162–171.

Salvati, L., Perini, L., Sabbi, A., & Bajocco, S. (2012). Climate aridity and land use changes: A regional-scale analysis. *Geographical Research, 50*(2), 193–203.

Salvati, L., & Zitti, M. (2008). Regional convergence of environmental variables: Empirical evidences from land degradation. *Ecological Economics, 68*(1–2), 162–168.

Salvati, L., & Zitti, M. (2009). Substitutability and weighting of ecological and economic indicators: Exploring the importance of various components of a synthetic index. *Ecological Economics, 68*(4), 1093–1099.

Satti, S. L., Farooq, A., Loganathan, N., & Shahbaz, M. (2014). Empirical evidence on the resource curse hypothesis in oil abundant economy. *Economic Modelling, 42*, 421–429.

Schoene, D. H., & Bernier, P. Y. (2012). Adapting forestry and forests to climate change: A challenge to change the paradigm. *Forest Policy and Economics, 24*, 12–19.

Seifollahi-Aghmiuni, S., Kalantari, Z., Egidi, G., Gaburova, L., & Salvati, L. (2022). Urbanisation-driven land degradation and socioeconomic challenges in peri-urban areas: Insights from Southern Europe. *Ambio, 51*(6), 1446–1458.

Shahbaz, M., Destek, M. A., Okumus, I., & Sinha, A. (2019). An empirical note on comparison between resource abundance and resource dependence in resource abundant countries. *Resources Policy, 60*, 47–55.

Sharma, S., & Henriques, I. (2005). Stakeholder influences on sustainability practices in the Canadian forest products industry. *Strategic Management Journal, 26*(2), 159–180.

Sheppard, J. P., Chamberlain, J., Agúndez, D., Bhattacharya, P., Chirwa, P. W., Gontcharov, A., & Mutke, S. (2020). Sustainable forest management beyond the timber-oriented status quo: Transitioning to co-production of timber and non-wood forest products—A global perspective. *Current Forestry Reports, 6*(1), 26–40.

Siry, J. P., Cubbage, F. W., & Ahmed, M. R. (2005). Sustainable forest management: Global trends and opportunities. *Forest Policy and Economics, 7*(4), 551–561.

Sivakumar, B. (2011). Global climate change and its impacts on water resources planning and management: Assessment and challenges. *Stochastic Environmental Research and Risk Assessment, 25*, 583–600.

Sloan, S., & Sayer, J. A. (2015). Forest resources assessment of 2015 shows positive global trends but forest loss and degradation persist in poor tropical countries. *Forest Ecology and Management, 352*, 134–145.

Smit, B., & Pilifosova, O. (2003). Adaptation to climate change in the context of sustainable development and equity. *Sustainable Development, 8*(9), 9.

Spies, T. A., Giesen, T. W., Swanson, F. J., Franklin, J. F., Lach, D., & Johnson, K. N. (2010). Climate change adaptation strategies for federal forests of the Pacific Northwest, USA: Ecological, policy, and socio-economic perspectives. *Landscape Ecology, 25*, 1185–1199.

Stern, N. (2008). The economics of climate change. *American Economic Review, 98*(2), 1–37.

Swamy, L., Drazen, E., Johnson, W. R., & Bukoski, J. J. (2018). The future of tropical forests under the United Nations sustainable development goals. *Journal of Sustainable Forestry, 37*(2), 221–256.

Taghizadeh-Hesary, F., Zakari, A., Alvarado, R., & Tawiah, V. (2022). The green bond market and its use for energy efficiency finance in Africa. *China Finance Review International, 12*(2), 241–260.

Tegegne, Y. T., Cramm, M., Van Brusselen, J., & Linhares-Juvenal, T. (2019). Forest concessions and the United Nations sustainable development goals: Potentials, challenges and ways forward. *Forests, 10*(1), 45.

Thanh, T. T., Van Song, N., Huyen, N. T. T., Ha, L. T., & Huong, T. T. L. (2024). An exploration of linkage between climate-related financial policies and natural rents: Evidence from the global database. *Resources Policy, 88*, Article 104450.

Tomao, A., Quatrini, V., Corona, P., Ferrara, A., Lafortezza, R., & Salvati, L. (2017). Resilient landscapes in Mediterranean urban areas: Understanding factors influencing forest trends. *Environmental Research, 156*, 1–9.

Vardopoulos, I., Tsilika, E., Sarantakou, E., Zorpas, A. A., Salvati, L., & Tsartas, P. (2021). An integrated SWOT-PESTLE-AHP model assessing sustainability in adaptive reuse projects. *Applied Sciences, 11*(15), 7134.

VijayaVenkataRaman, S., Iniyan, S., & Goic, R. (2012). A review of climate change, mitigation and adaptation. *Renewable and Sustainable Energy Reviews, 16*(1), 878–897.

Volodzkiene, L. (2025). 4 Gender Equality. *Societal Challenges and Opportunities of Low-Carbon Energy Transformations*, 139.

Wali, A., Alvira, D., Tallman, P. S., Ravikumar, A., & Macedo, M. O. (2017). A new approach to conservation: using community empowerment for sustainable well-being. *Ecology and Society*, *22*(4).

Wastl, C., Schunk, C., Leuchner, M., Pezzatti, G. B., & Menzel, A. (2012). Recent climate change: Long-term trends in meteorological forest fire danger in the Alps. *Agricultural and Forest Meteorology, 162*, 1–13.

Wilkie, M. L., Eckelmann, C. M., Laverdiere, M., & Mathias, A. (2002). Forests and forestry in small island developing states. *International Forestry Review, 4*(4), 257–267.

Willemen, L., Drakou, E. G., Dunbar, M. B., Mayaux, P., & Egoh, B. N. (2013). Safeguarding ecosystem services and livelihoods: Understanding the impact of conservation strategies on benefit flows to society. *Ecosystem Services, 4*, 95–103.

Wilson, G. A., Kelly, C. L., Briassoulis, H., Ferrara, A., Quaranta, G., Salvia, R., Detsis, V., Curfs, M., Cerda, A., El-Aich, A., Liu, H., Kosmas, C., Alados, C. L., Imeson, A., Landgrebe-Trinkunaite, R., Salvati, L., Naumann, S., Danwen, H., Iosifides, T., … Zhang, P. (2017). Social memory and the resilience of communities affected by land degradation. *Land Degradation and Development, 28*(2), 383–400.

Wunder, S. (2005). Macroeconomic change, competitiveness and timber production: A five-country comparison. *World Development, 33*(1), 65–86.

Xie, M., Irfan, M., Razzaq, A., & Dagar, V. (2022). Forest and mineral volatility and economic performance: Evidence from frequency domain causality approach for global data. *Resources Policy, 76*, Article 102685.

Yu, Y. (2023). Role of Natural resources rent on economic growth: Fresh empirical insight from selected developing economies. *Resources Policy, 81*, Article 103326.

Zahawi, R. A., Holl, K. D., Cole, R. J., & Reid, J. L. (2013). Testing applied nucleation as a strategy to facilitate tropical forest recovery. *Journal of Applied Ecology, 50*(1), 88–96.

Zahoor, Z., Latif, M. I., Khan, I., & Hou, F. (2022). Abundance of natural resources and environmental sustainability: The roles of manufacturing value-added, urbanization, and permanent cropland. *Environmental Science and Pollution Research, 29*(54), 82365–82378.

Zhang, Y., & Dilanchiev, A. (2022). Economic recovery, industrial structure and natural resource utilization efficiency in China: Effect on green economic recovery. *Resources Policy, 79*, Article 102958.

Zhang, Z., & Zhang, C. (2023). Revisiting the importance of forest rents, oil rents, green growth in economic performance of China: Employing time series methods. *Resources Policy, 80*, Article 103140.

Military Expenditure

9

Introduction

Starting from World War II, issues connected to military investments and expenditures have been of primary concern within frameworks dealing with international security, economic policy, and scholarly research (Akume & Akadiri, 2025). Within this context, the relationship between military spending and economic growth, as highlighted in a purely Keynesian view, took an increasing relevance (Cypher, 1987; Kentor & Kick, 2008; Seccareccia, 1995). According with this theory, government expenditures, including military investments, foster economic growth, especially during recession (Chary & Singh, 2024; Florişteanu, 2023; Odehnal & Neubauer, 2020). Conversely, the 'opportunity cost' theory highlights that economic resources connected to military provisions are likely to hinder the growth in sectors as health, education and infrastructure, in turn preventing (or slowing down) long-term development (Fourie, 2006).

At the same time, the significant levels of financial resources to develop ballistic, military, and energy technologies have deeply affected the outcomes of modern conflicts and the resulting geopolitical environment (Dunne et al., 2005). In general, the magnitude of these investments reflects a country's military potential but also measures depending on broader strategic priorities, national concerns and economic capabilities (Dumas, 2014). The second global conflict represents a crucial turning point, since the introduction of nuclear weapons radically challenged war strategies and international security paradigms (Shahbaz et al., 2013). The deployment of atomic bombs in Hiroshima and Nagasaki in 1945 not only remarked the destructive potential of technological advancements fostered by wartime expenditures but also underscored the compelling urgency of new policies of arm control (Gillani et al., 2022) and the implementation of global mechanisms to handle military issues effectively (Ikedinma, 2017).

This period saw exponential increases in defence budgets, driven by the desire to maintain strategic superiority, develop advanced weaponry, and project power globally (Njifen & Anemann, 2023). Arms' races resulted in a substantial allocation of financial resources toward nuclear and conventional weapon systems, often at the expense of socioeconomic development priorities (Adler, 1992; Blomberg, 1996; Ward, 1984). The Cold War period featuring an unprecedented arm race, primarily between the United States and the Soviet Union, saw an exponential increase in defence budgets often at the expense of investments to enhance economic development (Hove & Lillekvelland, 2018). This harmful evidence required a reliable data collection aimed at properly analysing military global trade trends (LaBrie et al., 2018). These occurrences led to the spread of several international organizations to promote peace and disarmament, such as the

C. Ciaschini and L. Salvati, *Sustainable Development*, Sustainable Development Goals Series,
https://doi.org/10.1007/978-3-032-12559-0_9

Stockholm International Peace Research Institute (SIPRI) collecting and analysing data on military expenditures (Archer, 2020; Beckley, 2010; Ripsman & Paul, 2005). The resulting evidence highlights the security priorities also identifying the tricky mechanisms that govern the relationship between military investments, economic development, and social stability (Adler & Greve, 2009; Fravel, 2011; Robinson, 2022). The compelling requirement of enhanced arms' supplies and defence consequent to recent outbreak of the Russia-Ukraine war, further highlights the relevance of monitoring military expenditure.

Within this framework, a renewed relevance acquires the indicator examining military spending as a percentage of the Gross Domestic Product (GDP). This measure depicts an almost exhaustive framework on the economic resources' allocation decisions of nations on security provisions. With reference to economic potential and the level of technological development at large, the indicator remarks reallocation priorities and potential trade-offs between such investments and those in areas as health, education, and social welfare (Greenwood, 2024). In addition, the indicator contributes to the achievement of Sustainable Development Goal 16 (SDG16), aimed at measuring and promoting peaceful and inclusive communities characterized by strong institutions (Bose & Kumar, 2021). It may also contribute to inform broader sustainability monitoring systems through indicators' dashboards (Salvati & Carlucci, 2014). In this perspective, a proper study of military expenditures could not demise from factors as national security, political aspects, lobbies and available financial resources.

Defence industries strongly affect government decisions on technological development, subtracting relevant financial resources to investments related to social and economic growth, e.g. healthcare, education and poverty alleviation, especially in developing countries (Milton, 2021; Neuman, 2010; Wang & Zhu, 2025). Within this context, policymakers are in charge of bringing together requirements connected to national security, and those involved in the achievement of peace and global stability (Maizels & Nissanke, 1986). They should handle the delicate security issues in light of the preservation of economic development (Calcara et al., 2023; Kiratli, 2022; Molas-Gallart, 2010). To this purpose, military investments, aimed at enhancing national security, should not undermine broader social and economic achievements (e.g. Wilson et al., 2017).

To this purpose, the work of organizations like SIPRI becomes crucial in studying global trends in military expenditures, shedding light also on the potential disarmament policies. Other institutions, as North Atlantic Treaty Organization (NATO), International Monetary Fund (IMF), and the United Nations (UN), have also establish some guidelines for military and paramilitary forces (Coleman et al., 2021). Nevertheless, major challenges still persist in the pursue of a standardized and comparable methodology quantifying military expenditures across countries, accounting for inhomogeneous definitions (Barnum et al., 2025), depending on country specific methodologies and bureaucratic traditions (Alptekin & Levine, 2012). In general, the inherent evolution in the global security landscape requires a proper and homogeneous global analysis of trends and motivations underlying military expenditures to foster a peaceful and sustainable world (Kentor & Jorgenson, 2017).

Moreover, the quantification of clandestine expenditures or intelligence activities involving secret arm programs are rather impossible (Heffington et al., 2019), further preventing a reliable provision of global data on the specific phenomenon (Kollias et al., 2018; Sedysheva, 2012; Smith, 2014). SIPRI military expenditure figures are presented on a calendar-year basis, assuming an even expenditure rate throughout the fiscal year. The aggregation method is weighted average (World Bank Indicators 2022). The original dataset retrieved in the World Bank consisted of 266 entries including countries and regional aggregates. Countries displaying missing values have been neglected in the present analysis.

The research mainly relies on the various annual bulletins published by SIPRI whose limits, with reference to data availability and reliability, have been extensively discussed in earlier literature. For instance, Collier et al. (2002) estimated the arms' races of neighbouring countries highlighting the strong impact of their expenditure in assessing the level of military expenditure of a certain country. Results have been presented in three parts. The first part describes maps and graphs generated and redrawn from the World Bank website. This part has been complemented with simplified plots delineating trends over time in the specific indicator. The second part performs a descriptive statistical analysis. The third part focuses on a Principal Component Analysis (PCA) of input data, properly assessing the relationships between the variables.

Literature Review

Literature on military expenditure highlights some worth mentioning issues, inherent specificities and peculiarities (Brzoska, 1981). For instance, although measurements began in 1960, data for many countries geopolitically crucial in the conflicts of the last 60 years have only been collected since the end of the Soviet Union (USSR), i.e. since the late 1980s. Data for some developing countries rely on incomplete databases or rough estimates (World Bank Indicators 2022). Despite the relevant efforts of SIPRI to ensure a full comparison of data series on military expenditure for each country over time, this aim is not always achieved for lacking uniform definitions and difficult verifications of the accuracy and completeness of input data (sensu Rontos et al., 2022). Data used for the study are expressed as a percentage of Gross Domestic Product (GDP). Collier et al. (2002) argued that governments determine the level of military expenditure based on those established by their neighbours. This interdependent behaviour may derive from a benevolent (e.g. emulation) or aggressive reason and, in many cases, is likely to trigger future conflicts (Navarro-Galera & Maturana, 2011).

Economic Drawbacks of Military Expenditures

The ratio between military expenditures and Gross Domestic Product (GDP) defines a relevant indicator adopted in the economic and geopolitical literature to investigate the strategic priorities and the defence capabilities of a country. This indicator, often described as a percentage of military expenditure over GDP, allows the evaluation of the economic commitment devoted to national security with reference to the overall size of the economy, providing a relative measure comparing countries characterized by different economies of scale. For a long time, literature has acknowledged that this measure represents not only a simple figure but also a complex indicator, affected by multiple economic, political, and strategic factors. One of the theoretical foundations of this analysis derives from the 'theory of public choices', according to which the allocation decisions of public resources, including the defence sector, result from negotiations between different political actors and group interests (Allio & Allio, 2020; Şimandan, 2011; Wu, 2008).

In this perspective, the possible 'trade-off' between military expenditure and other public investments (such as education, healthcare, or infrastructures) suggests that the increase in military expenditure could imply a reduction of available resources for another sector, with implications on social welfare and sustainable economic growth (Campbell & Kelly, 1994). In this sense, the ratio between military expenditure and GDP also acquires an interpretative role with respect to the political priorities of a government (Da Silveira & Slack, 2001). In other words, it represents a signal of commitment toward security or, on the contrary, of a possible biased allocation of public resources (Ahmadimousaabad et al., 2013).

Academic research based both on the 'economies of scale' assumption and on the 'network effects' perspective, suggests that increasing investments in defence could generate positive decreases in the industrial military sector, encouraging technological innovation, development of industrial capabilities, and job creation (Abel, 2018). Nevertheless, this relation is still object of debate within scholars, since the multiplicative effect of these investments tends to be limited throughout time and often characterized by inefficiencies, as emerged when analysing military expenditures of countries with ambiguous management systems (Lebovic, 1995).

A recurrent theme in literature relates to the cyclical dynamics and the security crisis affecting the ratio between military expenditure and GDP (Jakobsen & Halvorsen, 2018). After the Cold War, many developed countries reported a decrease in military expenditure over GDP, resulting from a reduction of international tensions and a changing strategic context (SIPRI, 2021). This tendency characterised an interpretative framework well-known with the name of the 'end of history', suggesting a slowdown of the conflicts between strong countries and a higher emphasis on international cooperation and economic development (Fukuyama, 2021; Huiqin, 2011; Tareke, 2021). Nonetheless, studies on non-transparent defence expenditures and parallel budgets, highlight the tendency of some countries to secretly increase their own military expenditures, due to regional crises and new asymmetric threats (Bing-Fu & Liming, 2006).

As remarked in the previous paragraph, a relevant issue in this kind of research is represented by the lacking reliability of data on military expenditures. Within this context, the recurrent practices of masking and manipulating data on expenditure prevent effective cross-country comparisons (Bachtiar, 2024; Fontanel, 2022; Waller, 1997). Alternative (integrated) techniques could be adopted to address the issue as the analysis of satellite images, monitoring the import of weapons, and statistical imputation, thus enabling a more accurate understanding of 'defence and security' dynamics (Anzoom et al., 2021).

Recently, the traditional concept of military expenditure has also been enriched and complicated with other nonconventional expenditures, e.g. cybersecurity, intelligence operations, and hybrid war (Brown, 2018). The resulting, more articulated, framework of the military capacity of a country is even more difficult of assess since these components, even crucial to evaluating the real military power and the deterrence potential of a country, do not emerge from official accounting schemes (Chin, 2023; Khan & Zeeshan, 2024; Wither, 2023). Generally, the economic and strategic literature highlights the relevance of the indicator to preserve political and economic stability (Alesina & Perotti, 1996). An excessive level of military expenditures could generate macroeconomic unbalances, inflation, and excessive debt, while low expenditure levels can compromise national security and the international credibility of a country (Voskanyan, 2022).

In this sense, the ratio also becomes an indicator of the 'sustainability' of defence policies, and its interpretation requires careful geopolitical and economic contextualization (Mia et al., 2025). In conclusion, the ratio between military expenditure and GDP represents a key element in the economic and geopolitical literature (Sweidan, 2023), and provides insights into the strategic priorities, available resources, and dynamics of international power (Blankenship & Miles Joyce, 2020).

Empirical Findings

At the year 1960, Syria, the United Kingdom (UK), and the United States (USA), are the only countries with the darkest colour gradation (their military expenditure exceeded 4.3% of GDP). At that time, indeed, the USA incurred in forced expenditure for the ongoing war in Vietnam; in Syria, there were riots and uprisings culminating in a failed *coup d'état* the following year. White areas reflect lacking data for most African states, together with China, the Soviet bloc, and the Arabian Peninsula. Many nations have in fact only recently entered the SIPRI database (for instance, data on Russia has only been officially collected by the Swedish institute since 1991, the year of the collapse of the USSR). In 2020, data on military expenditure were available for all nations except for South Sudan, Libya, Venezuela, Zimbabwe, Somalia, Yemen, Syria, Suriname, North Korea, Uzbekistan and Turkmenistan, mainly due to the conflicts plaguing these territories for years. Countries performing the highest military expenditure during Covid-19 were Saudi Arabia, Oman, Algeria, Morocco, Jordan, Israel, Azerbaijan and Armenia. In the latter two countries, these massive investments in weapons referred to the specific conflict in 'Nagorno Karabakh' region (SIPRI, 2021).

Over the period 1960–2020, the trend of military expenditure in the countries surveyed by SIPRI, highlights three periods of sharp increases. The first period refers to the mid-1960s and the early 1970s, characterised by the Cold War, the Vietnam War, the Yom Kippur conflict, and the related oil crisis. The second period, between 1982 and 1985, relates to the Falklands War between Argentina and the UK, and Israel's occupation of Southern Lebanon. In the last period, corresponding with the global financial crisis in 2009, a downturn was recorded compensated with positive values only in correspondence with Covid-19 (2020). This positive trend mainly relates to sudden increases in military expenditure of China and India due to the ongoing conflict between Pakistan and Kashmir and renewed border tensions with China (SIPRI 2020). Being fully in line with these empirical facts and findings, Fig. 9.1 illustrates a box–plot of the statistical distribution of military expenditure across world countries at the two years. The highest values, possibly representing 'geopolitical' or 'technical' outliers, seem to be concentrated in the first years of investigation. Wide fluctuations in the statistical distribution, however, emerge between 1968 and 1976, when the whiskers reach even negative values. Those years featured multiple wars especially in the Middle East, together with many tensions permeating most of the Western block due to the Cold War between the Soviet Union and the USA.

Descriptive Statistics

Table 9.1 shows the descriptive statistics of the analysed indicator, considering only the first year of each decade. The global mean value peaks in the 1970s, a sign that the tensions accumulated in the previous years have been finally contained. From that period onwards, however, the average fell steadily, rising again in 2020. The median value increased until 2.7 in 1980, further decreasing to values around 1.45 in both 2010 and 2020. All the values of median over mean ratio are slightly lower than 1, indicating a left–skewed statistical distribution of military expenditures across countries. Skewness dynamics (sensu Salvati & Zitti, 2009) easily confirmed such assumptions.

All values of kurtosis, even if decreasing, are greater than 0, thus indicating a leptokurtic distribution. These two latter indicators suggest the presence of outliers that make the distribution strongly skewed (Egidi et al., 2020), particularly in the first years of the survey. A percentile anal-

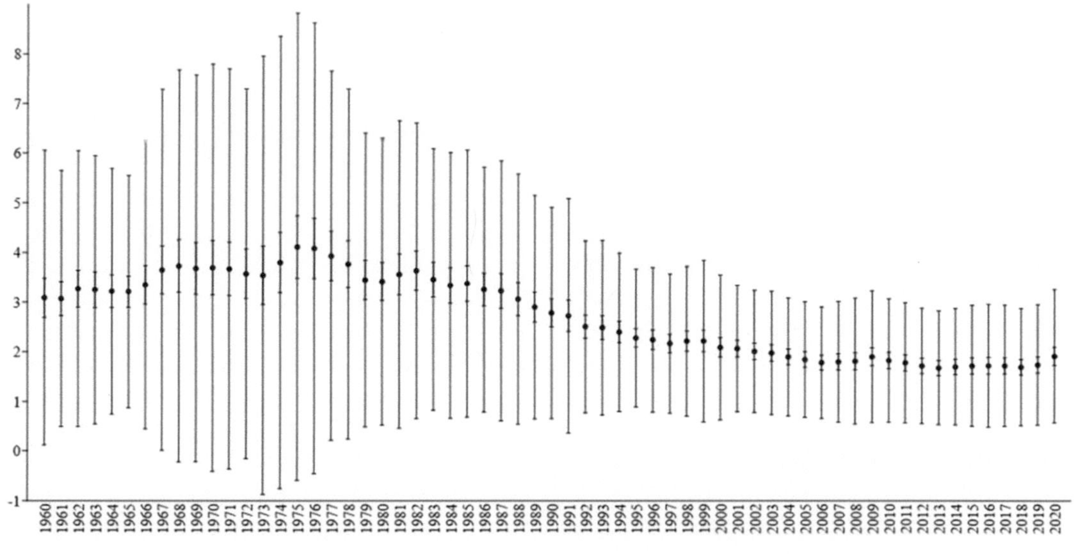

Fig. 9.1 A graphical summary of the statistical distribution of military expenditures over GDP rate across world countries based on a mean and whisker plot. *Source* Authors' elaboration of World Bank indicators

Table 9.1 Descriptive statistics summarizing the statistical distribution of military expenditures across world countries

Year	1960	1970	1980	1990	2000	2010	2020
Mean	3.09	3.69	3.41	2.78	2.09	1.83	1.91
Min	0.10	0.50	0.40	0.30	0.40	0.40	0.30
Max	20.40	23.40	18.90	12.40	7.60	5.90	6.70
Norm. range	6.58	6.20	5.40	4.35	3.45	3.01	3.35
Median	2.45	2.40	2.70	2.35	1.70	1.45	1.45
Median/mean	0.79	0.65	0.79	0.85	0.81	0.79	0.76
25° percentile	1.63	1.90	1.80	1.63	1.20	1.03	1.00
75° percentile	3.48	3.50	4.20	3.08	2.50	2.45	2.68
Per(75–25)/median	0.76	0.67	0.89	0.62	0.76	0.98	1.16
Skewness	3.97	3.44	3.25	2.64	2.01	1.61	1.52
Kurtosis	21.02	12.79	14.72	8.72	4.46	2.88	2.58

Source Authors' elaboration of World Bank indicators

ysis depicts an almost linear statistical distribution for both years (1960 and 2020), with rising values only around the 90th percentile, possibly connected to outliers affecting the entire distribution. In other words, the graph documents that 10% of the countries bear military expenditures above 5% of GDP in 1960 and above 4% of GDP in 2020. Considering the last percentile fractions, only a low percentage of countries had expenditures in 1960 greater than 8% of GDP.

Correlation Analysis

Linear (Pearson) and nonlinear (Spearman) correlation coefficients (Fig. 9.2) display almost similar values (Salvati & Carlucci, 2011), except some divergences between the two lines observed in correspondence with the years having the highest military expenditure between 1961 and the first half of the 1980s. Comparable and high correlation coefficients indicate a substantially stable development path against time (Salvati & Zitti, 2008). Outliers include Jordan (with a peak in 1960), Israel and the United States.

Multidimensional Analysis

The results of a Principal Component Analysis (PCA) run on standardized input data, suggest that military expenditure is significantly determined by macro-level factors (PC1, explaining 76% of the total variance), with marginal influences of PC2 and PC3 (respectively, accounting for 10% and 6% of total variance). PC1 dominance outlines the existence of a single leading pattern in military expenditure, likely affected by few significant factors (e.g. Gross Domestic Product, population size and density, or regional conflicts). To this purpose, Fig. 9.3 illustrates a PCA biplot depicting the relationship between countries, i.e. dots, and arrows, reflecting the trend of military expenditure as a percentage of GDP over years.

Along the two main axes, arrows highlight a rather stable trend of military expenditure, even with short times of increased expenditure due to geopolitical events and regional conflicts. However, while most countries align closely, specific time periods perform deviations affected by 'outlier' behaviours. Countries projected near the central cluster, generally exhibit trends in line with the average ones, as a result of their peculiar geopolitical, economic, or historical contexts. For instance, Tunisia (TUN) aligns with global and regional averages while exhibiting slightly higher (or temporarily reduced) expenditures due to socio-political factors (e.g. the 'Arab spring').

Austria (AUT) and Norway (NOR) are stable European countries, likely reflecting NATO and regional expenditure standards, while Ecuador (ECU) reflects Latin America's trends of modest military investment. Outliers, e.g. Israel and Egypt, significantly deviate from the main cluster,

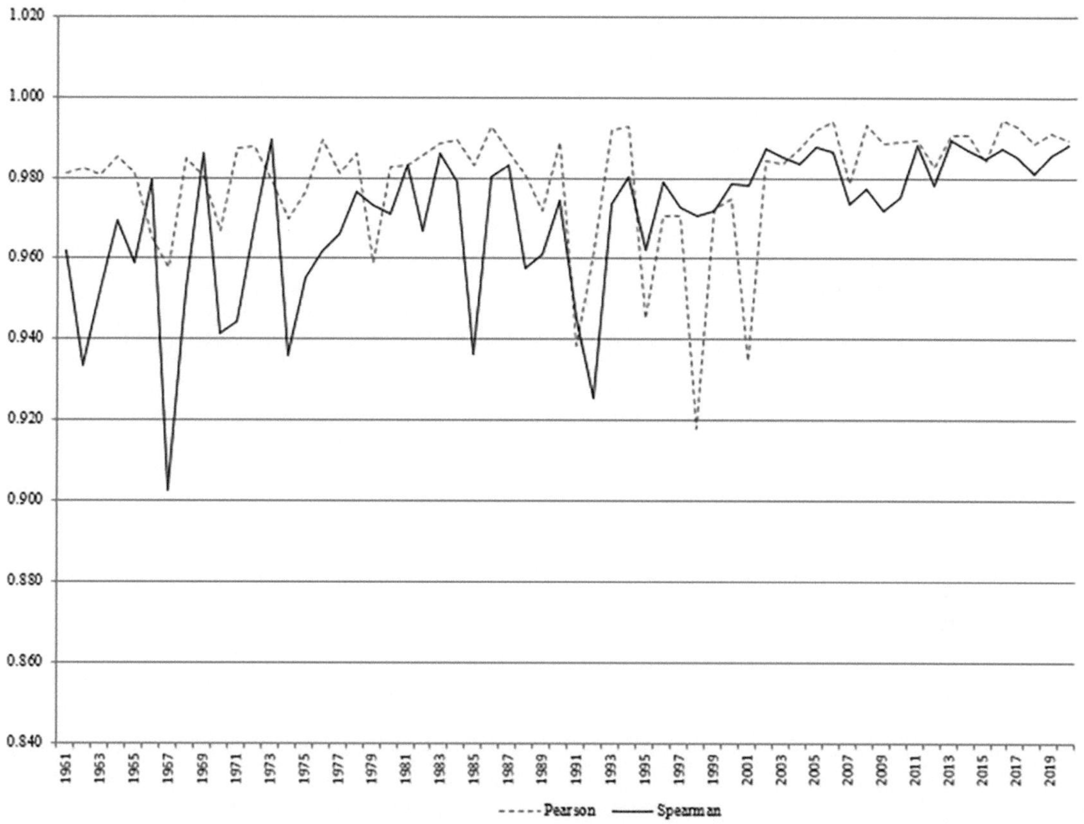

Fig. 9.2 Trends over time in parametric (Pearson) and nonparametric (Spearman) correlation coefficients within the (−1) lagged time series of military expenditure over GDP ratio, by year. *Source* Authors' elaboration of World Bank indicators

depicting unique military expenditure patterns in the area. Israel (ISR) is located on the positive side of Component 1, performing higher military expenditure driven by primary factors as economic scale, geopolitical influence, and regional security challenges. The country is, indeed, well known for the high level of investments in military sector relative to GDP due to persistent security concerns and a robust military infrastructure. Egypt (EGY) was projected on the negative side, suggesting distinct trends in military expenditure compared with other countries. This country allocates a lower proportion of resources to military expenditure compared to countries clustered on the positive side of Component 1 since its economic constraints still limit defence budgets.

Discussion

This study illustrated long-term trends in the amount of military expenditure incurred by each country as a percentage of GDP, quantifying the share of income and wealth allocated for weapons. Descriptive statistics document most of the expenditure as concentrated in the first decades of observation, i.e. from 1960 to the mid-1980s. In the 1960s, the nuclear arms race within the conflict between the American and Soviet bloc justified such expenditure regime (Brender & Drazen, 2013). A slowdown in arms' supply was observed after the fall of the Berlin Wall (Albalate et al., 2012). Over the last twenty years, relevant increases in military expenditure concentrated in countries of the 'Global South' (Bove et al., 2017).

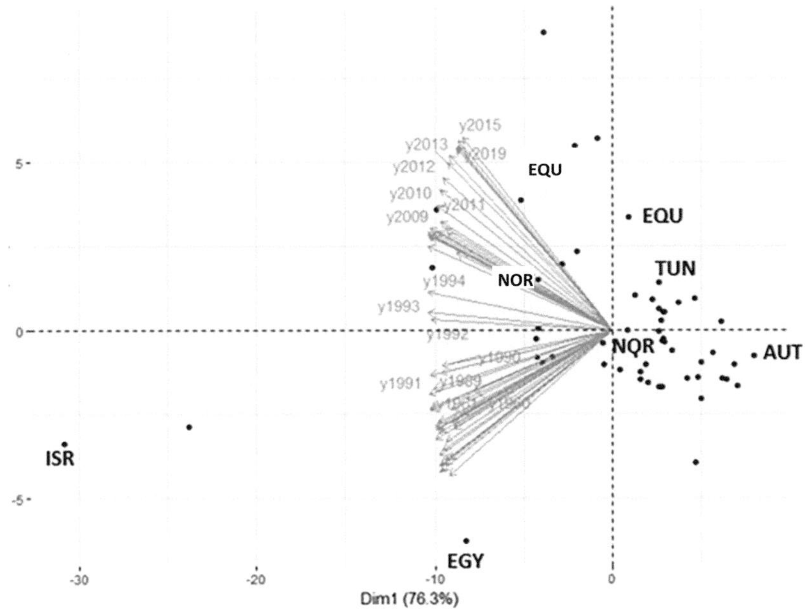

Fig. 9.3 Biplot of the first two components from PCA. *Source* Authors' elaboration of World Bank indicators

Empirical evidence from PCA highlights a similar pattern over the years for most Western countries (Powell et al., 2018). The empirical findings collected in this study are basically in line with the 'rearmament between neighbours' hypothesis' (Collier et al., 2002) basing the rearming of a country on the decisions of its neighbours. At the same time, the main increases in percentage military expenditure over GDP seem to coincide with years of change in the internal political order, as the election of a new president in a given country. Nevertheless, results should also account for the lacking data on SIPRI database especially for developing countries.

In general, the work attempts to provide suggestions for future policy decisions based on a proper analysis of the already existing measures (Lamonica et al., 2020). In its 2020 Yearbook, SIPRI reports several geopolitical upheavals in Covid-19 years, along with a relevant increase in weapons purchases (Markusen, 2003). As already remarked, the analysis lacks some data for 2021 and 2022. Nevertheless, an increased rearmament expenditure is retrieved across the borders. In particular, it is worth mentioning the huge amounts of arms sent by the European Union and the USA to Ukraine, as well as the latent conflict in Afghanistan—where the Taliban regained power in August 2021—or the Israeli-Palestinian conflict. Last but not least, lobbies heavily affect national governments in collecting funds for weapons, even outside the military context (Liff & Ikenberry, 2014).

Concluding Remarks

The inherent complexity in official statistics of military expenditure, and the key role of public budgets in any sustainable development strategy in both advanced and emerging economies, claims for a multidisciplinary approach, securing reliable indicators and official statistics, while integrating quantitative analysis, strategic considerations, and political evaluations, to effectively contribute to the formulation of policies of sustainable security and liability in a hopefully global peaceful environment.

References

Abel, A. B. (2018). Optimal debt and profitability in the trade-off theory. *The Journal of Finance, 73*(1), 95–143.

Adler, E. (1992). The emergence of cooperation: National epistemic communities and the international evolution of the idea of nuclear arms control. *International Organization, 46*(1), 101–145.

Adler, E., & Greve, P. (2009). When security community meets balance of power: Overlapping regional mechanisms of security governance. *Review of International Studies, 35*(S1), 59–84.

Ahmadimousaabad, A., Bajuri, N., Jahanzeb, A., Karami, M., & Rehman, S. (2013). Trade-off theory, pecking order theory and market timing theory: A comprehensive review of capital structure theories. *International Journal of Management and Commerce Innovations, 1*(1), 11–18.

Akume, M. A., & Akadiri, S. S. (2025). Unpacking the impact of military spending on economic growth in resource-rich nations. *International Journal of Economic Policy Studies*, 1–24.

Albalate, D., Bel, G., & Elias, F. (2012). Institutional determinants of military spending. *Journal of Comparative Economics, 40*(2), 279–290.

Alesina, A., & Perotti, R. (1996). Income distribution, political instability, and investment. *European Economic Review, 40*(6), 1203–1228.

Allio, R., & Allio, R. (2020). Economic globalization, realpolitik, new wars. *War in Economic Theories over Time: Assessing the True Economic, Social and Political Costs*, 217–244.

Alptekin, A., & Levine, P. (2012). Military expenditure and economic growth: A meta-analysis. *European Journal of Political Economy, 28*(4), 636–650.

Anzoom, R., Nagi, R., & Vogiatzis, C. (2021). A review of research in illicit supply-chain networks and new directions to thwart them. *IISE Transactions, 54*(2), 134–158.

Archer, C. (2020). Peace movement work on military spending. In *Military Spending and Global Security* (pp. 118–133). Routledge, London.

Bachtiar, M., Ahmad, I., Sahabuddin, Z. A., & Trismadi, T. (2024). A Comprehensive meta-analysis of determinants influencing military expenditure: New methodological insights and implications for defence budget allocation. *Statistics, Politics and Policy.*

Barnum, M., Fariss, C. J., Markowitz, J. N., & Morales, G. (2025). Measuring arms: Introducing the global military spending dataset. *Journal of Conflict Resolution, 69*(2–3), 540–567.

Beckley, M. (2010). Economic development and military effectiveness. *The Journal of Strategic Studies, 33*(1), 43–79.

Bing-Fu, C., & Liming, Z. (2006). The determinants of China's defense expenditure before and after transition. *Conflict Management and Peace Science, 23*(3), 227–244.

Blankenship, B., & Miles Joyce, R. (2020). Purchasing power: US overseas defense spending and military statecraft. *Journal of Conflict Resolution, 64*(2–3), 545–573.

Blomberg, S. B. (1996). Growth, political instability and the defence burden. *Economica*, 649–672.

Bose, S., & Kumar, A. (2021). Conflict-affected areas and sustainable development goal 16. *Peace, Justice and Strong Institutions* (pp. 120–130). Springer International Publishing.

Bove, V., Efthyvoulou, G., & Navas, A. (2017). Political cycles in public expenditure: Butter versus guns. *Journal of Comparative Economics, 45*(3), 582–604.

Brender, A., & Drazen, A. (2013). Elections, leaders, and the composition of government spending. *Journal of Public Economics, 97*, 18–31.

Brown, J. (2018). An alternative war: The development, impact, and legality of hybrid warfare conducted by the nation state. *Journal of Global Faultlines, 5*(1–2), 58–82.

Brzoska, M. (1981). The reporting of military expenditures. *Journal of Peace Research, 18*(3), 261–275.

Calcara, A., Gilli, A., & Gilli, M. (2023). Short-term readiness, long-term innovation: The European defence industry in turbulent times. *Defence Studies, 23*(4), 626–643.

Campbell, D. E., & Kelly, J. S. (1994). Trade-off theory. *The American Economic Review, 84*(2), 422–426.

Chary, S., & Singh, N. (2024). On the role of military spending: An economic thought perspective. *Cogent Social Sciences, 10*(1), 2350837.

Chin, W. (2023). The western military vision of future war. In *War, Technology and the State* (pp. 87–114). Bristol University Press, Bristol.

Coleman, K. P., Lundgren, M., & Oksamytna, K. (2021). Slow progress on UN rapid deployment: The pitfalls of policy paradigms in international organizations. *International Studies Review, 23*(3), 455–483.

Collier, P., & Hoeffler, A. (2002). Military expenditure: Threats, aid, and arms races. *Aid, and Arms Races (November 2002).*

Cypher, J. M. (1987). Military spending, technical change, and economic growth: A disguised form of industrial policy? *Journal of Economic Issues, 21*(1), 33–59.

Da Silveira, G., & Slack, N. (2001). Exploring the trade-off concept. *International Journal of Operations and Production Management, 21*(7), 949–964.

Dumas, L. J. (2014). The real effects of military spending on security. *Peace Economics, Peace Science and Public Policy, 20*(3), 377–393.

Dunne, J. P., Smith, R. P., & Willenbockel, D. (2005). Models of military expenditure and growth: A critical review. *Defence and Peace Economics, 16*(6), 449–461.

Egidi, G., Salvati, L., & Vinci, S. (2020). The long way to tipperary: City size and worldwide urban population trends, 1950–2030. *Sustainable Cities and Society, 60*, Article 102148.

Florişteanu, E. (2023). Military and defence expenditures in the current security context. Concepts and influencing factors. *Scientific Bulletin-Nicolae Balcescu Land Forces Academy, 28*(2), 194–204.

Fontanel, J. (2022). Is the SIPRI estimate of military expenditure a reliable indicator of the power of states? *Defence Economics.*

Fourie, J. (2006). Economic infrastructure: A review of definitions, theory and empirics. *South African Journal of Economics, 74*(3), 530–556.

Fravel, M. T. (2011). Economic growth, regime insecurity, and military strategy: Explaining the rise of noncombat operations in China. *Asian Security, 7*(3), 177–200.

Fukuyama, F. (2021). *After the end of history: Conversations with Francis Fukuyama*. Georgetown University Press.

Gillani, S., Ahmed, A., Khan, A. N., & Hussain, A. (2022). Military expenditures, income inequality and economic growth interlinkages: An empirical assessment for developing countries. *Review of Applied Management and Social Sciences, 5*(2), 109–128.

Greenwood, D. (2024). Defence and national priorities since 1945. In *British Defence Policy in a Changing World* (pp. 174–207). Routledge, London.

Heffington, C., Park, B. B., & Williams, L. K. (2019). The "Most Important Problem" dataset (MIPD): A new dataset on American issue importance. *Conflict Management and Peace Science, 36*(3), 312–335.

Hove, K., & Lillekvelland, T. (2018). Investment cost escalation–an overview of the literature and revised estimates. *Defence Inflation*, 37–59.

Huiqin, H. (2011). Ideological choices in contemporary China: With a discussion of Francis Fukuyama's declaration of the end of history. *International Critical Thought, 1*(2), 158–170.

Ikedinma, H. A. (2017). Impact of military spending on african development. *Ife Social Sciences Review, 25*(1), 98–111.

Jakobsen, J., & Halvorsen, T. (2018). The Durability of the security dilemma: An empirical investigation of action-reaction dynamics in states' military spending (1988–2014). *The Chinese Journal of International Politics, 11*(2), 153–192.

Kentor, J., & Kick, E. (2008). Bringing the military back in: Military expenditures and economic growth 1990 to 2003. *Journal of World-Systems Research*, 142–172.

Kentor, J., & Jorgenson, A. (2017). Military expenditures and health: A cross-national study, 1975–2000. *International Journal of Sociology and Social Policy, 37*(13/14), 755–772.

Khan, B., R., & Zeeshan, F. (2024). Gap between conventional and non-conventional threats: understanding and addressing modern security challenges. *International Journal of Policy Studies, 4*(2).

Kiratli, O. S. (2022). Together or not? Dynamics of public attitudes on UN and NATO. *Political Studies, 70*(2), 259–280.

Kollias, C., Paleologou, S. M., Tzeremes, P., & Tzeremes, N. (2018). The demand for military spending in Latin American countries. *Latin American Economic Review, 27*, 1–17.

LaBrie, R. C., Steinke, G. H., Li, X., & Cazier, J. A. (2018). Big data analytics sentiment: US-China reaction to data collection by business and government. *Technological Forecasting and Social Change, 130*, 45–55.

Lamonica, G. R., Recchioni, M. C., Chelli, F. M., & Salvati, L. (2020). The efficiency of the cross-entropy method when estimating the technical coefficients of input–output tables. *Spatial Economic Analysis, 15*(1), 62–91.

Lebovic, J. H. (1995). How organizations learn: US government estimates of foreign military spending. *American Journal of Political Science*, 835–863.

Liff, A. P., & Ikenberry, G. J. (2014). Racing toward tragedy? China's rise, military competition in the Asia Pacific, and the security dilemma. *International Security, 39*(2), 52–91.

Maizels, A., & Nissanke, M. K. (1986). The determinants of military expenditures in developing countries. *World Development, 14*(9), 1125–1140.

Markusen, A. R. (2003). The case against privatizing national security. *Governance, 16*(4), 471–501.

Mia, R., Shakil, N. A. F., & Ahmed, I. (2025). A conceptual architecture for proactive global cyber defense: Exploiting technical, adversarial, and geostrategic dimensions. *International Journal of Advanced Cybersecurity Systems, Technologies, and Applications, 9*(1), 21–50.

Milton, S. (2021). Higher education and sustainable development goal 16 in fragile and conflict-affected contexts. *Higher Education, 81*(1), 89–108.

Molas-Gallart, J. (2010). Innovation, defence and security. In *The Theory and Practice of Innovation Policy*. Edward Elgar, Chichester.

Navarro-Galera, A., & Maturana, R. I. O. (2011). Innovating in defence policy through spending efficiency: The life cycle costing model. *Journal of Policy Modeling, 33*(3), 407–425.

Neuman, S. G. (2010). Power, influence, and hierarchy: Defense industries in a unipolar world. *Defence and Peace Economics, 21*(1), 105–134.

Njifen, I., & Anemann, A. (2023). Military expenditures and human capital development in sub-Saharan Africa: A system GMM approach. *Development Studies Research, 10*(1), 2163678.

Odehnal, J., & Neubauer, J. (2020). Economic, security, and political determinants of military spending in NATO countries. *Defence and Peace Economics, 31*(5), 517–531.

Powell, J., Faulkner, C., Dean, W., & Romano, K. (2018). Give them toys? Military allocations and regime stability in transitional democracies. *Democratization, 25*(7), 1153–1172.

Ripsman, N. M., & Paul, T. V. (2005). Globalization and the national security state: A framework for analysis. *International Studies Review, 7*(2), 199–227.

Robinson, P. (2022). The Russia-Ukraine conflict and the (Un) changing character of War. *Journal of Military and Strategic Studies, 22*(2).

Rontos, K., Ermini, B., & Salvati, L. (2022). Enlarging the divide? Per-Capita Income as a measure of social inequalities in a southern European City. *Quality & Quantity*, 1–17.

Salvati, L., & Carlucci, M. (2011). The economic and environmental performances of rural districts in Italy: Are competitiveness and sustainability compatible targets? *Ecological Economics, 70*(12), 2446–2453.

Salvati, L., & Carlucci, M. (2014). A composite index of sustainable development at the local scale: Italy as a case study. *Ecological Indicators, 43*, 162–171.

Salvati, L., & Zitti, M. (2008). Regional convergence of environmental variables: Empirical evidences from land degradation. *Ecological Economics, 68*(1–2), 162–168.

Salvati, L., & Zitti, M. (2009). Substitutability and weighting of ecological and economic indicators: Exploring the importance of various components of a synthetic index. *Ecological Economics, 68*(4), 1093–1099.

Seccareccia, M. (1995). Keynesianism and public investment: A left-Keynesian perspective on the role of government expenditures and Debt. *Studies in Political Economy, 46*(1), 43–78.

Sedysheva, M. (2012). Strategic management system and methods of controlling as key elements of military expenditure policy-making process. *Journal of Strategy and Management, 5*(3), 353–368.

Shahbaz, M., Afza, T., & Shabbir, M. S. (2013). Does defence spending impede economic growth? Cointegration and causality analysis for Pakistan. *Defence and Peace Economics, 24*(2), 105–120.

Şimandan, M. (2011). Recent developments in the theory of markets with imperfect information. *Journal of Economics and Business Research, 17*(1), 107–126.

SIPRI (2021), SIPRI Yearbook 2021: Armaments, disarmament and international security. Stockholm International Peace Research Institute.

Smith, R. P. (2014). The economic costs of military conflict. *Journal of Peace Research, 51*(2), 245–256.

Sweidan, O. D. (2023). Geopolitical risk and military expenditures: Evidence from the US economy. *Russian Journal of Economics, 9*(2), 201–218.

Tareke, S. G. (2021). From the end of history to the end of neo-liberalism: From Fukuyama to Fukuyama. *African Journal of Political Science and International Relations, 15*(4), 139–147.

Voskanyan, M. A. (2022). Government policy to maintain macroeconomic stability: The case of Armenia. *Journal of New Economy, 23*(1), 24–42.

Waller, D. (1997). Estimating non-transparent military expenditures: The case of China (PRC). *Defence and Peace Economics, 8*(2), 225–241.

Wang, S., & Zhu, X. (2025). Defence technology management: The effect of the manufacturing base on arms exports. *International Journal of Technology Management, 97*(4), 500–516.

Ward, M. D. (1984). Differential paths to parity: A study of the contemporary arms race. *American Political Science Review, 78*(2), 297–317.

Wilson, G. A., Kelly, C. L., Briassoulis, H., Ferrara, A., Quaranta, G., Salvia, R., Detsis, V., Curfs, M., Cerda, A., El-Aich, A., Liu, H., Kosmas, C., Alados, C. L., Imeson, A., Landgrebe-Trinkunaite, R., Salvati, L., Naumann, S., Danwen, H., Iosifides, T., … Zhang, P. (2017). Social memory and the resilience of communities affected by land degradation. *Land Degradation and Development, 28*(2), 383–400.

Wither, J. K. (2023). Hybrid warfare revisited: A battle of 'Buzzwords'. *Connections: The Quarterly Journal, 22*(1), 7–27.

Wu, J. Y. (2008). A general behavior model and new definitions of organizational cultures. *The Journal of Socio-Economics, 37*(6), 2535–2545.

Index

C. Ciaschini and L. Salvati, *Sustainable Development*, Sustainable Development Goals Series, https://doi.org/10.1007/978-3-032-12559-0

P

R

S

W

Y

MIX
Papier aus verantwortungsvollen Quellen
Paper from responsible sources
FSC® C105338

If you have any concerns about our products,
you can contact us on
ProductSafety@springernature.com

In case Publisher is established outside the EU,
the EU authorized representative is:
Springer Nature Customer Service Center GmbH
Europaplatz 3, 69115 Heidelberg, Germany

Printed by Libri Plureos GmbH
in Hamburg, Germany